Collective Learning for Transformational Change

This book offers a step by step guide for those seeking to undertake a transformational change process based on strong collaboration among diverse interests. Guiding transformational change goes beyond small changes to an existing system. It leads to lasting change in the system itself. The collective learning process achieves a systems change through a continuous learning spiral based on open learning among diverse interests. The sixteen case studies cover guided transformational change in personal learning, team-building, community development, organizational change, monitoring and evaluation, and cross-cultural learning. Each transformational change has been treated as a celebration of mutual learning.

Part one of the book provides an outline of the theory and practice of collective learning. The theory draws on the experiential learning cycle developed by David Kolb. The practice follows the rules of open space learning, dialogue and valuing diversity. The case studies in Part two are examples of collective learning leading to transformational change in a wide range of contexts, from cities to councils to organisations. Part three offers thirty-three activities on which the programme designers can draw in the course of guiding transformational change.

This innovative book will be of immense value to researchers, students and professionals in the fields of organizational change, organizational behaviour, management education, and sustainability training, education and leadership.

Valerie A. Brown is Director of the Local Sustainability Project, Human Ecology program, Fenner School of Environment and Society, The Australian National University.

Judith A. Lambert is Director of Community Solutions, Sydney, Australia and works in the interface between the social and environmental aspects of sustainable living.

Collective Learning for Transformational Change

A guide to collaborative action

Valerie A. Brown and Judith A. Lambert

First published 2013
by Routledge
2 Park Square, Milton Park, Abingdon, Oxon, OX14 4RN

Simultaneously published in the USA and Canada
by Routledge
711 Third Avenue, New York, NY 10017

Routledge is an imprint of the Taylor & Francis Group, an informa business

© 2013 Valerie A. Brown and Judith A. Lambert

The right of Valerie A. Brown and Judith A. Lambert to be identified as author of this work has been asserted by them in accordance with sections 77 and 78 of the Copyright, Designs and Patents Act 1988.

First issued in paperback 2013

All rights reserved. The purchase of this copyright material confers the right on the purchasing institution to photocopy pages which bear the photocopy icon and copyright line at the bottom of the page. No other parts of this book may be reprinted or reproduced or utilised in any form or by any electronic, mechanical, or other means, now known or hereafter invented, including photocopying and recording, or in any information storage or retrieval system, without permission in writing from the publishers.

Trademark notice: Product or corporate names may be trademarks or registered trademarks, and are used only for identification and explanation without intent to infringe.

British Library Cataloguing in Publication Data
A catalogue record for this book is available from the British Library

Library of Congress Cataloging-in-Publication Data
Brown, Valerie A.
 Collective learning for transformational change : a guide to collaborative action / Valerie A. Brown and Judith A. Lambert.
 p. cm.
 Includes bibliographical references and index.
 1. Social change. 2. Organizational change. 3. Organizational learning.
 I. Lambert, Judith A. II. Title.
 HM831.B76 2013
 303.4–dc23
 2012017774

ISBN13: 978-0-415-62292-9 (hbk)
ISBN13: 978-0-203-10567-2 (ebk)
ISBN13: 978-0-415-82621-1 (pbk)

Typeset in Univers by Clive Hilliker

Contents

List of Figures	viii
List of Tables	ix
List of Boxes	x
Sally and Richard	x
Foreword	xi
Acknowledgements	xv

PART 1. INSTRUCTIONS
Collective Learning for Transformational Change — 1

1.	The theory: Collective social learning	3
2.	The practice: Party time	21
3.	Following the collective learning spiral	31
4.	Step 1. Setting the scene: Who to invite?	39
5.	Step 2. Collective ideals: What should be?	49
6.	Step 3. Collective facts: What is?	53
7.	Step 4. Collective ideas: What could be?	57
8.	Step 5. Collective action: What can be?	63
9.	Step 6. Following on	67
10.	Guiding transformational change	71

PART 2. CASE STUDIES
Celebrations of Collective Learning   87

11.	Holding the party	89
12.	Managing whole-of-community change: Bon voyage	97
13.	Introducing new ideas: A cocktail party	115
14.	Initiating long-term change: Opening night	125
15.	Changing problem communities: Housewarming	147
16.	Achieving collective thinking: Coming of age	155
17.	Monitoring and evaluation: Street party	165
18.	Teamwork: Bring a plate	175
19.	Working from the Guidebook: Going it alone	193
20.	Summing up	215

PART 3. RESOURCES
A–Z of Collective Learning   223

Introduction	225
Adaptive management	229
Alliancing	230
Balancing the players	231
Collaboration	233
Conflict resolution	234
Consultation	236
Conversation	238
Dialogue	240
Event management	241
Forecasting	243
Gatekeepers	244
Hosting	245

Imagining	246
Joining in	247
Knowledge brokering	248
Learning styles	249
Multiple knowledges	250
Negotiation	251
Open Space Technology	252
Pattern languages	253
Problem-solving games	254
Questioning	256
Risk and risk-taking	257
Synergy and synthesis	258
Team-building	259
Transdisciplinarity	261
Understanding	262
Values mapping	263
Visioning	265
Wicked problems	267
Xing the minefield	268
Yarning	269
Zany ideas	270

BIBLIOGRAPHY 271

INDEX 275

Figures

1.1	The key elements of experiential learning	7
1.2	Kolb's learning styles and learning stages	10
2.1	Multiple knowledges in each of the four stages of collective learning	22
3.1	Four stages of the collective learning cycle = one turn of a spiral	31
3.2	Three turns of the collective learning spiral	35
4.1	The parties involved in lasting transformational change	40
4.2	The knowledges needed for transformational change	41
4.3	Individual positions on transformational change	42
4.4	The mandala of collective learning	43
4.5	Knowledge cultures' contributions to a collective decision	46
4.6	Knowledge cultures can hear each other	47
LC-1	Learning cycle stage 1: developing ideals	49
LC-2	Learning cycle stage 2: describing the facts	53
6.1	Field force map	54
LC-3	Learning cycle stage 3: designing ideas	57
7.1	Staying ahead of the game	59
LC-4	Learning cycle stage 4: taking action	63
10.1	Differences between solving simple and complex problems	72
10.2	Kolb learning styles linked to a range of occupations	76
10.3	Attitudes to change	77
10.4	Classic gestalt figure: 'Christ in the snow'	79
10.5	Model of a PhD thesis based on collective learning spiral	81
10.6	Questionnaire for participants in a collective learning spiral	82
10.7	Interaction between perspectives on learning	84
11.1	Finding the whole picture	90
12.1	Collective learning cycle for sustainability and health project	99
12.2	Sample postcard from Art of Moving project	104

12.3	Sample of the fifty Art of Moving postcards on replacing cars with physical exercise	105
14.1	Ideals for a collective learning alliance	140
14.2a	Affinity circle	143
14.2b	Priority connections formed by Alliance members	143
16.1	Steve Strike photo of Uluru	161
16.2	Sally Morgan painting of Uluru	163
17.1	The parallel learning cycles of a collective knowledge initiative and its monitoring and evaluation	168
17.2	The collective learning cycle as a monitoring tool	172
17.3	Community sustainability goals	173
17.4	Symbols for community sustainability goals	174
18.1	Book cover	176
18.2	The five braided strands of social learning	188
19.1	Raw data as collected at the Montessori workshop	199
19.2	Collective social learning process for workshop	206
20.1	All together now!	221
A–Z.1	Building collective learning and action	226
A.1	Standard adaptive management cycle	229
B.1	Keirsey's four major temperaments	232
C.1	The dynamics of conflict resolution	235
C.2	Arnstein's ladder of public participation	236

Tables

11.1	What sort of celebration fits which sort of change?	92
14.1	Grid cross-referencing multiple knowledge needs and resources	142
14.2	Sample of personal evaluations of collective learning experiences	145
18.1	Design of writers' workshop and book commitments	191

Boxes

1.1	Seven Ages of Man	4
1.2	A Matter of Time and Place	6
1.3	Action Learning	16
1.4	Stages of Collective Learning	17
7.1	The Collective Learning Buzz	58
10.1	Einstein as a Collective Thinker	78
12.1	Report of an International Review Committee	108
14.1	A Poem Written after a Community Meeting	133
16.1	Building Collective Knowledge as a Lateral Thinker	158
16.2	A Linear Thinker as a Collective Thinker	160
17.1	Key to Figure 17.1 in Practice	169
18.1	A Value Line	181
18.2	Book Outline as Allocated to Author Groups	183
18.3	Book Outline	189
19.1	Correspondence Between Workshop Organizer and Possible Consultant	195

Sally and Richard

1	Meetings, bloody meetings	25
2	A leap into new territory	30
3	My goodness! It worked!	38
4	Now for the big one	48
5	A roller-coaster ride	61
6	We got there!	65
7	Walking the talk	94
8	Keeping a learning journal	134
9	Are we there yet? Reflecting on a shared journey	216

Foreword

Val and Judy have produced an important Guide for individuals, specialists and decision-makers desiring transformational change. It is important because it provides mechanisms for bringing those parties together, while at the same time bringing together respective strands comprising sound theoretical underpinnings; real world good practice examples; clear exercises to try out and essential tools for collective work. These rich resources, together in the one package, will have strong appeal for practitioners.

It is an important Guide because it enables practitioners to creatively confront blockages to transformational change – blockages which are described in the text as

- The ongoing separation between social, environmental and economic concerns
- Competition rather than collaboration between communities, experts and governments
- Disconnects between local, regional and national scales
- An apparently increasingly fragmented society

It is an important Guide because of its offer of hope to those tackling 'wicked problems' (defined as those that can only be restored by changes in the society that generated them) together with new concepts, approaches and tools to address these. How refreshing it is to be introduced to thinking about transformational change of society's wicked problems that embrace paradox, collaboration and ethics. And while primed for outcomes, an acceptance of the potential of no final solution is likely to break new ground for many conditioned or driven to needing to achieve a solution at any cost!

It is an important Guide because it describes approaches for applying learning creatively and together!

Part 1

Part 1 outlines theoretical concepts with a deliciously light touch. It takes readers, with ease, through explanations of collective learning, learning styles and how individuals apply these. It describes how switching between styles does not come naturally, but that knowing about and having access to the styles (as in a team) can develop the capacity to use them all for transformational change. It outlines an approach which seeks a common language to involve all interests equally and values all types of knowledge.

Part 1 also describes how deliberative transformational change needs a constructive, ethical direction which takes account of the aims and diversity of all parties, celebrates the future possibilities of combined and diverse interests and guards watchfully against negativity. This is a direction which is designed to strengthen a community as a whole; contribute to understanding about how the world works and allow for individual and whole of organizational learning.

It describes how the application and testing of models has uncovered new issues for consideration. These include:

- Reflecting on the necessity of sequence or ordering in the guidance of transformational change
- Observing the constraints of modelling in our Western cultural context and the need to understand different cultural contexts
- Understanding that the nature of knowledge is not just individually–focused, but also applies to communities (leading to community development); specialist knowledge (understood as research design) and organizational knowledge (expressed as strategic plans)
- Ground-truthing (which is what it says!)

What further exciting advances can we therefore look forward to as a result of the adoption (or adaptation) of this modelling?

Some years back, I had a conversation with a friend, who admitted she was challenged by change. As we both worked for a community organization advocating social change, I somewhat tongue-in-cheek suggested she should 'embrace change and be excited by it!'. Her response gave me pause for thought: 'Oh but I am, as long as change does not happen to me!'

Part 2

Part 2 of the Guide, therefore, appeals to my sense that transformational change is worthy of embrace and should be exciting as well as instructive. It provides readers with grounded case studies, narration of successful collective learning and exercises to guide and trial actions. We are invited to consider these as types of parties or celebrations to which we are invited or from which we can select, to effectively become our own party-planner. I love this concept of celebration of collective learning. It should appeal at a fundamental level to our party culture and sense of fun (transformational change should not be a drudge). At the same time, this concept is delivering on serious benefits: stepping readers through learning stages, presenting opportunities to learn from difference and enabling a change of 'leadership' styles from individual to collective interests.

Part 3

Part 3 of the Guide provides readers with a treasure trove of tools for transformational change. 'Why are tools vital?' the Guide asks. Could we not assume that as adults we know what we are supposed to do? It is suggested that we are probably all so attuned to individual or organizational learning styles that we can each benefit from some remedial action to work collectively. I would add that new 'tips and tricks' are always welcome to practitioners. Some of the tools in the A–Z list will not be new to practitioners, others will surprise and delight. It includes everything from Alliancing to Negotiation, Open Space Technology to Zany Ideas and how to plumb these. And a final word for readers – based on previous use, in many different community settings - these tools work, and are fun to apply. Enjoy!

Jackie Ohlin, Consultant,
Public Policy, Urbis
Australia Asia Middle East

January 2012

Acknowledgements

This Guidebook has emerged from our work with literally hundreds of agents of change and thousands of their participants in collective learning across Australia and Asia. We have selected examples for their variety and for illustrations of a particular principle, not because some were better than others. All were inspiring.

In preparing the Guide, sometimes we have transferred ideas from one case study to another so that a point can be made. Sometimes we have condensed wonderful projects to enable us to fit them in. Sometimes there have been political change processes and personal tensions which it is not wise or helpful to document in published form. Therefore we have used pseudonyms so that no-one is embarrassed or compromised.

We, the authors, are entirely responsible for the final form of the book. However, the Guide contains the inspired work of many individuals, of whom we can name but a few.
We have to particularly thank:

- Greg Bruce
- Susan Butler
- Niki Carey
- Judy Charnaud
- Erica Fisher
- Nigel Grier
- Wilma Grier
- John Harris
- Kerryn Hopkins
- Heather Pearce
- Geoffrey Pryor
- Wendy Rainbird
- Tracey Rich
- Skye Rose
- Kellie Walters

for contributing their work, ideas and support for this Guide.

Artwork:

Sally Morgan's painting of Uluru by kind permission of the Penguin Group (Australia).

Steve Strike's Desert Storm photograph by kind permission of Outback Photographics.

Michael Leunig's cartoon by generous personal permission of Michael Leunig.

Art of Moving postcards by design students at the University of Canberra by kind permission of the Canberra Environment and Sustainability Resource Centre.

Cartoons by David Pope by kind permission of Local Sustainability Project, The Australian National University.

Poem by Melinda Hillery by permission of the author.

Book design and facilitation: Clive Hilliker.
We greatly appreciate the professional skill and collaborative style with which Clive Hilliker facilitated the text design, layout and diagrams for this book. It has been a joy to work with him as he developed our ideas into an accessible format.

We are also grateful to our publishers Earthscan and particularly Khanam Virjee and Helena Hurd for their interest and advice throughout the project.

Valerie Brown and Judy Lambert
Canberra, Australia, April 2012

PART 1. INSTRUCTIONS

Collective Learning for Transformational Change

1 The theory: Collective social learning

Summary: This chapter begins by exploring social learning. It then discusses David Kolb's work on individual experiential learning. His ideas are extended into a collective social learning capable of guiding transformational change.

Note: This chapter is for those who wish to know the theory behind the practice. For readers with a greater interest in the practical aspects, rather than the theory, go straight to Chapter 2. Here you'll find a template for guiding transformational change. You can then choose from the 16 case studies in Part 2. Read those along with the template to guide your collective learning process.

About social learning

Social learning is part of our everyday life. It is the pathway through which we learn to live in a shared world, a world that will inevitably be different from that of our parents and different again for our children. This is an era of continuous rapid social and environmental change. Apparently simple changes in social behaviours, such as an increase in everyday use of childcare centres, or the ready take-up of mobile phones, are but symptoms of core changes in work patterns and family units, which in turn change our interaction with both our social and our physical environments.

Personally and professionally, we are all by definition involved in social learning. It has made us who we are, and allows us to fit into the society in which we were reared. Social learning inevitably goes beyond that of each individual to shape the whole of society. The effectiveness of a society's capacity to change is marked by the willingness of the members of the society to go beyond their traditional social practices. Transformational social change involves questioning existing rules and boundaries, and finding new solutions and ways of living. However, rather than being welcomed, change can be strongly resisted. The pull of the traditional ways of defining individual goals, professional practices, and organizational cultures can be stronger than the push of the need to change. Yet changes in all of these are needed if there is to be a transformational change (Box 1.1).

> **Box 1.1 Seven Ages of Man**
>
> In the sixteenth century Shakespeare described seven ages of 'man': infant, schoolboy, lover, soldier, judge, aging, second childhood. At that time the use of the word 'man' was accurate since the ages applied only to the males of his society, when only boys went to school or became citizens. Today a social transformation has meant each of the transformational ages applies to both sexes.
>
> Every society has traditions that mark each of the seven age transformations by collective social learning. For instance, in 21st century Western society coming-of-age from child to adult is marked by physiological changes, voting, driving, drinking, and earning. Each of these actions involves the society's structural supports: politics, law, health, science and economics. There is no such thing as an entirely free individual; we are all created by our society. Even a protester recognizes the social rules – they just want to change them.

Social learning can therefore be a mixed blessing. On the one hand it is the glue that holds society together, a cultural inheritance that is passed down through the generations, and the security of knowing that things will go on as they always have. On the other hand traditional social learning can act as a brake on change, since it is more concerned with maintaining the old than introducing the new. The interaction between the traditionalists and the innovators is therefore a significant tension. In times of transformational change the tension can escalate into outright conflict. Hence the political unrest sweeping around the planet, and the impasse between responses to climate change, well-documented in Les Brown's 'Vital Signs: the trends that are shaping our future'.

With constant transformational change becoming a distinctive feature of our time, some conflict is inevitable. The pressures for change manifest themselves in many ways. The impact of human activities on the planet is so great that changes are necessary for a viable human future. Financial crises follow one another with no end in sight. Advances in technology appear to offer the answers, and then the answers cause further disruption. Communities need to support their members through the rapid pace of social change. Scientists, politicians, industry leaders and communities each offer solutions, but they are competing solutions. The question is, how can we move forward to welcome change within this chaos?

A first step is to recognize the need for a social learning that celebrates rather than impedes change – a redirected social learning. A second step is to recognize that major change necessarily generates complex social issues. These complex issues require a quite different approach to that of addressing the simple problems of maintaining business as usual. The American wit, Mencken, writes 'For every complex problem there is a solution which is simple, clean and wrong'.

A simple problem can be solved through applying the simple logic of cause and effect; a complex problem asks for a far more comprehensive understanding of the issues.

For instance, a troubled community may need improved social services, or it may need to develop new mutual support systems to cope with change. Halting environmental degradation may mean protecting a threatened species, or it may mean developing a comprehensive environmental management system involving all parties. If new technology is to provide more than a stop-gap solution, it will require complementary changes throughout a society. In each of these examples, the second more comprehensive change will require mutual learning and understanding across all the sectors of the society.

The Theory: Collective Social Learning   5

The capacity for mutual learning among all the interests in a society may well be the key to successful transformational change. It is not so easy to achieve, however, in a society with strong divisions between ways of thinking about the world. The divisions between ages, gender, beliefs and values, places, and income levels build strong walls of thought and language which serve to strongly reinforce the existing system. To bring change to one is to threaten the continuity of the others. The entry of women into the workforce, increasing economic inequality and an extended life expectancy are but a few of the changes affecting Western social systems. Learning is needed to bring windows into those walls, so that changes can be judged on their full effect on the whole of society.

Box 1.2 A Matter of Time and Place

What constitutes politics, law, health, science and economics is a matter of time and place. To the ancient Greeks achieving democracy meant individual citizens speaking in the market place, in our era a formal process involving over a billion people globally every three or four years. The delivery of nineteenth-century law was punitive in the extreme, now there is an emphasis on potential rehabilitation. Health is very much a matter of time and place: life expectancy in Europe rose from forty to eighty years in a century, sharply altering the expectations of Shakespeare's ages.

Transformational changes are therefore part of the human condition. A major difference between then and now is a rising confidence in the human capacity for deliberate human-directed transformational change. In past times, transformations have been regarded as the business of the gods, of authoritative figures, and of chance. There has been a twentieth-century social transformation from accepting the world as it is, to the belief that humans could intentionally reconstruct their own world. Dramatic changes in the planetary environment resulting from human activities, such as the increase in the ozone hole and rise in mean temperatures, have confirmed this belief. The practice of social learning has been extended to guided transformational change.

David Kolb and experiential learning

Since all learning starts in the individual human head, individual learning is a good place to start before considering social learning as a group exercise. A seminal writer on individual adult learning, David Kolb, offers a blueprint for individual learning that can be used across a whole society to guide harmonious transformational change. Building on the work of other influential educators, the Kolb learning cycle draws on Jean Piaget's stages of intellectual development, Paulo Freire's emphasis on the need for learning through conscious reflection, Kurt Lewin's insights into organizational cultures and Carl Jung's assertion that learning styles result from people's preferred ways of adapting in the world.

Kolb put forward the proposition that all individual learning is based in the learner's reflection on direct experience. His approach is therefore called experiential learning. In its simplest form experiential learning is a matter of completing a cycle, first feeling that something is important, then watching what is actually happening, thinking about the possible consequences and acting on those thoughts (Figure 1.1).

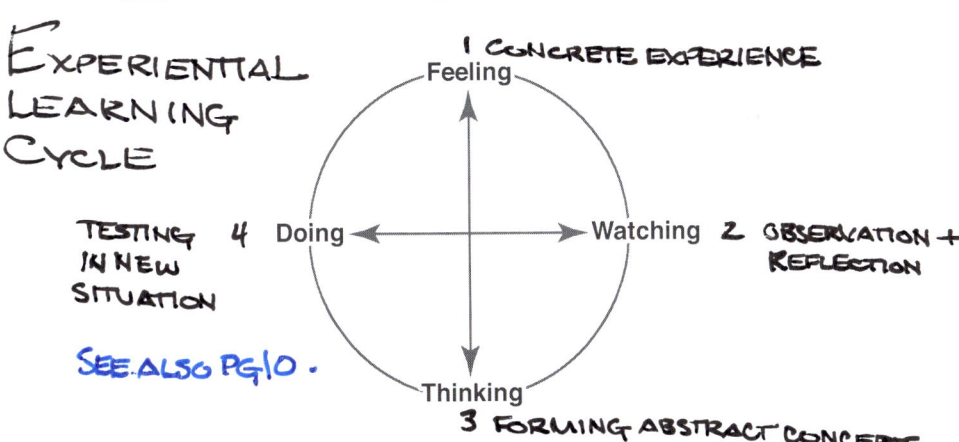

Figure 1.1 The key elements of experiential learning

David Kolb expressed these four forms of experience as concrete experience, observation and reflection, forming abstract concepts and testing them in new situations. He represented these in the famous experiential learning circle in a sequence that involves, for any one issue,

1. **Feeling:** ideals developed from past experience, followed by

2. **Watching:** facts determined by observations, followed by

3. **Thinking:** new ideas generated by using the imagination, followed by

4. **Doing:** actions that test the new ideas in the old situation (Figure 1.2).

The changed situations offer new experiences that challenge previous values and so the cycle becomes a spiral. Kolb and his colleagues found that the same cycle appears time and again as the basis for any substantial and lasting learning.

Two aspects of the learning spiral are especially noteworthy: the use of concrete, 'here-and-now' experience to test new ideas; and the use of feedback from different aspects of experience to change existing practices and theories (Kolb 1984: 21–22). Kolb named his model experiential learning to emphasize the link with the work of John Dewey, Paulo Freire and Jean Piaget, all of whom stress the role direct experience plays in learning.

Observations of the learning cycle made by many researchers across a wide range of situations confirmed that the learning cycle could be generalized to all significant adult learning. The observations also confirmed that different interests gave a different emphasis on different parts of the learning cycle.

In our divided social system, standard social learning has ensured that learners tend to have already developed a strength in, or orientation to, one dimension of the learning cycle. Scientists choose to prioritize observation, planners depend on their capacity to think strategically, caring services draw on an ability to feel compassion and practical occupations learn by doing. These differences are explored in Chapter 10 on guiding transformational change.

After decades of studies in the field, researchers found that there were some consistent learning styles associated with certain personalities and occupations. A learning style inventory based on a self-report attitude scale was found to place people in four basic learning styles associated with the four stages of experiential learning. Kolb's model is particularly important in that it offers both a way to understand individual people's different learning styles, and also an explanation of a cycle of experiential learning that applies to us all.

8   Collective Learning for Transformational Change

Kolb writes that ideally (although not always) the learning process follows a learning cycle or spiral where the learner 'touches all the bases', i.e. the cycle of experiencing, reflecting, thinking, and acting. The influence of a particular personality or a particular set of social learning leads the learner to emphasize one or more of the stages and so to develop their own learning style. Kolb's model therefore works on two levels – a four-stage cycle and a four-type definition of associated learning styles.

Kolb observed that we cannot do or watch (practical actions) and think or feel (emotional actions) at the same time. Our urge to want to do both creates conflict, which we resolve through choice when confronted with a new learning situation. We internally decide whether we wish to watch or do, or whether to think or feel. The result of these two decisions helps to form throughout our lives a preferred learning style.

The four learning styles represented combinations of two preferred styles, rather like a two-by-two matrix of the four-stage cycle styles (Figure 1.2) for which Kolb used the terms diverging, assimilating, converging and accommodating.

Descriptions of the **four Kolb learning styles** are adapted from Smith's (2001) encyclopaedia of informal education:

1. **diverging (feeling and watching)** - These people are able to look at things from different perspectives. They prefer to watch rather than do, tending to gather information and use imagination to solve problems. They are best at viewing concrete situations from several different viewpoints. Diverging people perform better in situations that require ideas-generation, for example, brainstorming.

2. **assimilating (watching and thinking)** - The assimilating learning preference is for a concise, logical approach. Ideas and concepts are more important than people. These people require good clear explanation rather than practical opportunity. They excel at understanding wide-ranging information and organizing it in a clear logical format. People with this style are more attracted to logically sound theories than approaches based on practical value.

3. **converging (doing and thinking)** - People with a converging learning style can solve problems and use their learning to find solutions to practical issues. They prefer technical tasks, and are less concerned with people and interpersonal aspects. People with a converging learning style are best at finding practical uses for ideas and theories. They make decisions by finding answers to questions.

4. • **accommodating (doing and feeling)** - The accommodating learning style is 'hands-on', and relies on intuition rather than logic. These people build on other people's analysis, and link these to an experiential approach. They are attracted to new challenges and experiences, and to carrying out plans. They commonly act on 'gut' instinct rather than logical analysis.

As with any behavioural model, this is a guide not a set of rules. Anyone can use any learning style, and everyone has the potential to access all the styles. Nevertheless most people clearly exhibit strong preferences for a given learning style. The ability to use or 'switch between' different styles is not one that we should assume comes easily or naturally to many people. Special conditions are needed to encourage them to do so. People who prefer the 'assimilating' learning style will not be comfortable being thrown in at the deep end without notes and instructions. People who prefer to use an 'accommodating' learning style are likely to become frustrated if they are forced to read lots of instructions and rules, and are unable to apply their learning through experience as soon as possible.

Since collective learning requires contributions from different learning styles at different stages of the cycle, any deliberate transformational change requires access to all those styles. This may be in the form of a team in which each style is represented, or it may involve the interest groups being willing to learn to develop the capacity to use all the styles.

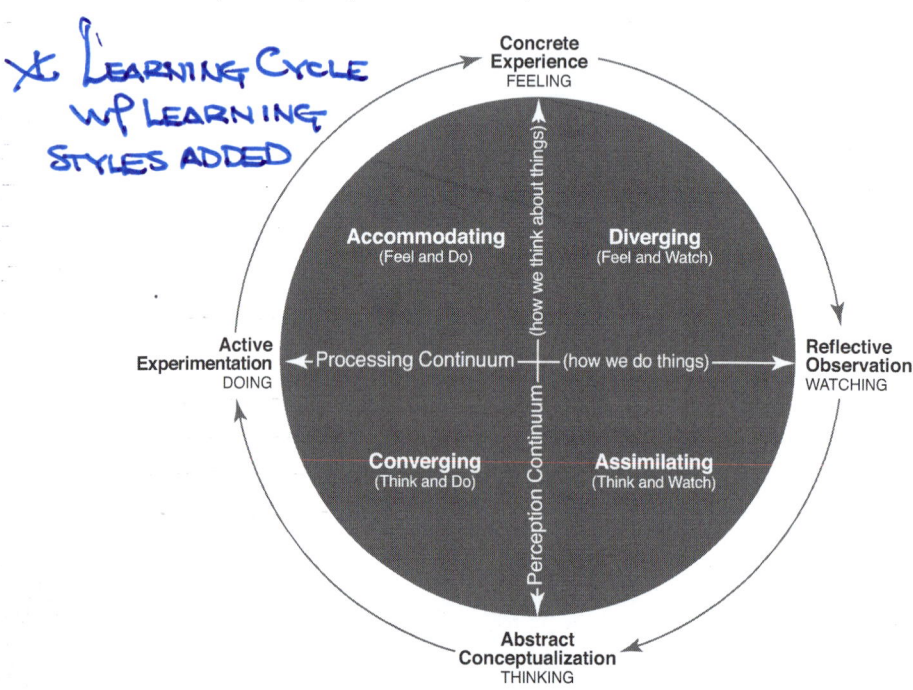

Figure 1.2 Kolb's learning styles and learning stages (adapted from Chapman and Calhoun, 2006)

Kolb explains how different people develop their own learning style. Many factors, including social setting, past experience, personality and occupation go to developing a person's preferred style. Kolb suggests that our ability to reconcile and successfully integrate the four different learning styles improves as we mature through the development stages established by the particular social learning environment. The development stages that Kolb identified are:

1. **Acquisition** - Birth to adolescence - development of basic abilities and 'cognitive structures' along biologically determined lines of development from egocentric to abstract thought.

2. **Specialization** - Schooling, early work and personal experiences of adulthood - the development of a particular 'specialized learning style' shaped by 'social, educational, and organizational socialization'. Specialization appears to be a feature of all cultures, strongly emphasized in Western approaches to knowledge.

3. **Integration** - Mid-career through to later life - expression of non-dominant learning styles in work and personal life. Lived experience appears to soften the socially-learnt emphasis on specialization. Hence the phrase 'wisdom of the elders', a feature of most cultures that has been sharply reduced in Western cultures.

Knowing a person's (and your own) learning style enables learning to be oriented across the potential spectrum and applied collectively at each learning stage. That said, everyone responds to and needs the stimulus of all types of learning styles to one extent or another – it's a matter of using emphasis that fits best with the given situation and a person's learning style preferences. Tools which allow the collective learning facilitator or guide to identify their own and their colleagues' learning styles can be found in Chapter 10 and in the tools in Part 3 of this Guidebook.

Among many other correlations between definitions, Kolb points out that Jung's 'Extraversion/Introversion' dialectical dimension (which features and is measured in the Myers-Briggs Type Indicator) correlates with the 'Active/Reflective' (doing/watching) dialectic (east-west continuum) of Kolb's model. Also, the Myers-Briggs 'Feeling/Thinking' dimension correlates with the Kolb model Concrete Experience/Abstract Conceptualization dimension (north-south continuum, Figure 1.2).

Like any model, the Kolb experiential learning stages are notable for what they do not do, as well as what they do. In applying the Kolb individual learning model to collective learning the following limitations of the Kolb model are taken into account:

The Theory: Collective Social Learning

Reflection: the Kolb model does not pay sufficient attention to the links between the stages, which require time for and a definite process of reflection.

Order: Kolb does not apply the stages in a fixed order. In developing collective learning it became clear that there is an important order when guiding transformational change.

Culture: The learning styles and stages are Western culturally based. There are different cultural models of identity which differ from the 'western' assumptions that underpin the Kolb model.

Knowledge: While David Kolb agrees that learning and knowledge are intimately related he doesn't really explore the nature of knowledge in any depth. He approaches the structure of knowledge from a social psychology perspective and his focus is upon informed, committed action by the individual.

Collective learning and transformational change

We have discussed the ways in which social learning can be conservative or innovative, forward or backward looking, mainstream or deviant. Intentional transformation will need to be innovative and forward-looking. It may be mainstream or deviant, depending on the level of acceptance in the social environment. No matter what the orientation, it will always involve bringing diverse individuals together for whole-of-community change. A fifteen year study by the Local Sustainability Project of The Australian National University explored the use of the Kolb learning cycle as the basis for collective social learning in collaborative projects with members of a wide range of communities.

In practice, applying collective learning to social change was found to require building on and extending the individual-based Kolb cycle. The strengths are the recognition of the close relationship between social learning and social change, establishing the four learning stages as a prototype for all adult learning, and making the connections between emotions and actions, facts and values. The extensions are from individual to collective learning, from a single interest to combining multiple interests, from simple to complex problem-solving and from serving society to changing society.

The risks of using the collective learning cycle without a clear direction include the possibility of a group using the cycle for social engineering for personal gain, and the potential for directionless drifting. To ensure that the change is guided towards a constructive ethical direction, there is need for a focusing question that takes account of the aims of all the parties.

12 Collective Learning for Transformational Change

In working across Australia and in Europe, Canada and the Pacific, the Local Sustainability Project members collaborated with hundreds of local communities. The sixteen case studies that make up Part 2 have been chosen from these collaborations to demonstrate the range of possible approaches to transformational change. The lessons for those intending to guide a deliberate transformational change are explored in Chapter 10. The theoretical basis for the collective learning cycle follows here, in the spirit of:

> *'Everything comes from everything, and everything is made from everything, and everything can be turned into everything else.'*
> Leonardo da Vinci, Notebooks

Once the individual experiential framework is extended to collective experiential learning it becomes apparent that this is already in widespread use among the contributing interests. Each of the key interest groups needed for whole-of-community change, that is, individuals, community, specialists, organizations and holistic thinkers, rely on their own version of the learning cycle for learning within their own group.

Kolb concentrates on the ways in which individuals learn. Yet each of the interest groups collectively follows the stages of the Kolb learning cycle, developing ideals, describing the facts, designing from new ideas, and doing the ideas in action (see Figures 2.1 and 3.1). For an individual, this is the basis for personal learning, while for a community it is called community development. For a specialist, it is a research design, for an organization a strategic plan. For an holist it is the transformative vision. In seeking transformational change, the cycle becomes a spiral. Each cycle of learning builds on the learning from the preceding cycle to respond to or to create further change.

In community development, community educators such as Malcolm Knowles describe the Kolb learning stages as:

1. Start where the learner is at: developing ideals for what should be

2. Undertake a reality check: describing concrete experiences, what is

3. Introduce the new: designing to optimize ideas of what could be

4. Evaluate old and new together: actively experimenting by doing in practice.

The goal is to strengthen the community as a whole. Vehicles for the change may be any of the social services, education, and self-help. But a community cannot change through its own internal resources alone. It requires other influential groups to be involved in the change.

A standard scientific inquiry brings about change through choosing a topic, developing a hypothesis about its importance (feeling), describing the topic in detail (watching), designing the possibilities (thinking) and testing the possibilities in action (doing). The results confirm or deny the hypothesis and so generate a fresh learning cycle within that field of science. The findings can bring transformational change in understanding how the world works: think of Newton and gravity, Mendel and genetics, Einstein and relativity.

Organizational change is directed by an organization's own agenda. Whether the organization is government, non-government or industry, its strategic planning for the future follows a familiar cycle: to develop goals, describe the setting, design a plan, put the plan into action. In standard practice, the goals are the objectives of the organization. In times of change the goals will involve a change of agenda and so require whole-of-organization learning. Unfortunately, much strategic planning leaves out the first and last stages of the cycle. All the action remains with the planners, and no change actually takes place. Many plans lie gathering dust on office shelves for lack of commitment to action.

Artists have been called 'the early warning systems of mankind'. The contribution of the arts to transformational change can be dramatic. Picasso's Guernica was painted to alert the world to the horror of the first aerial bombing of a civilian population. The French impressionist painters opened the eyes of a generation to their own environment. The photograph of a small girl aflame from a napalm bomb helped end the twentieth century Vietnam war. In each case the artist has had a flash of inspiration (feeling), considered what they wish to say (watching), decided how it could be represented (thinking), and then executed the work of art (doing). A work of art changes the society into which it is released and then other artists take up the challenge of the new.

Therefore collective learning has to find ways to bring together five versions of the experiential learning cycle. Each version applies collective learning for transformational change within its own group. Since whole-of-community change involves all five groups, the challenge is to find a common language and a common process that involves all of the interests equally. Trialling the Kolb process in many different contexts led to a form of the cycle which included individuals, community, specialists, organizations, and the arts or other holistic thinkers in guided transformational change.

In a collective learning cycle, the set of interest groups join together to answer a series of four questions which focus on the change they wish to make (Box 1.4).

Collective Learning for Transformational Change

A. Question 1. Stage 1 of the learning cycle: feeling and developing ideals. **What should be?** This question asks for the ideals of each participant in the change enterprise. The answer reveals the contributor's values. Individuals will answer with personal goals, community members with local visions of the future, specialists with the goals of their specialized interest and organizational representatives with their hoped for outcomes. Hopefully, creative thinkers will offer ideas which extend the goals of the entire group. This initial start to the collective learning allows each interest group to hear and respect the ideals of others.

B. Question 2. Stage 2 of the learning cycle: watching and describing the facts. **What is?** This stage involves all the interests in contributing the facts supporting and impeding the change. This will entail five different perspectives on just what are the facts. Individuals will contribute the facts derived from their own experience, communities from remembered shared events; specialists from a particular disciplinary framework; organizations from their business plan and the arts or other holists from imagination. It is important to be aware of and compensate for a power hierarchy that unfortunately in reality exists among the interest groups, facts from specialist knowledge are valued more than the strategic understanding of organizations, and both more than the facts contributed by communities, individuals or the creative thinkers.

C. Question 3. Stage 3 of the learning cycle: thinking and designing fresh ideas for change. **What could be?** This is the time for bringing the ideas of all the interests together in a synergy, that is, the contributors working together create something better than any could contribute alone. Therefore it is essential that each interest group makes its own contribution. A moment's thought establishes that any overarching idea about how to bring about whole-of-community change will need all the interests joined in a mutual brainstorm, the more diverse the interests the better.

D. Question 4. Stage 4 of the learning cycle: ideas into action through doing. **What can be?** Following Kolb, the fresh ideas generated collectively by all the interests need to be put into practice before any ongoing learning can take place. Collaborative action is therefore the essential final step. Collaboration itself can be a challenge for the different interests in the change – the key individuals, the affected community, the influential organizations, and the creative thinkers. Thus this final learning stage will need to draw on the many forms of collaboration being developed among the interests themselves: community self-determination, adaptive management, multi-focal organizations and applied design.

For each individual, the continuing spiral represents not just one episode, but life-long learning. Kolb established that only at the completion of a cycle where the learning has been tested in action does it become actionable knowledge. Box 1.3 offers an opportunity to test this from personal experience.

About collective learning

Memorable examples of collective social change highlight how whole-of-society learning may build up slowly, but the actual change can occur rapidly as all the interests come together in a significant event. President John F. Kennedy's assassination confirmed that the world was not a safe place. Changes in core symbols, such as the symbolic fall of the Berlin Wall between East and West Germany, or Copernicus and Galileo placing the sun, not Earth, at the centre of the universe, deliver a vision of a different community. While the moment of change can be initiated from one significant event, there has been a previous build-up of ideas. The transformational change was the combination of them all. Less dramatic but still confirming the human capacity for change are the sixteen case studies of celebrating change to be found in Part 2.

Kolb's open learning cycle offers a synergistic framework that is not a strait-jacket. Moreover, it follows plain common sense. It also allows each interest group to expand its own version of reality, while working collectively towards some agreed purpose (Box 1.3).

[Handwritten annotation: LEARNING ONLY OCCURS WHEN FULL CYCLE]

Box 1.3 Action Learning

Consider the last time you went to an inspiring lecture or workshop. What happened afterwards? Did you return to work or home intending to put the new ideas into practice? If you did try to put them into practice, what happened? And if you didn't, what happened then?

In the first case, when you tried to introduce the new ideas, were they welcomed? Did people listen to you? Without some experiential learning by the other people, it is very unlikely the ideas will be adopted, or welcomed or even listened to.

In the second case, if you went back to your previous setting and didn't try out the ideas, did you find that a week later you could remember only that you had had an interesting time, but not the ideas themselves?

Each of these common experiences confirm Kolb's conclusion that learning only takes place if the full cycle of ideals, facts, ideas and actions is completed.

Kolb and his colleagues and the Local Sustainability Project teams found that, in any given interest group, one of the four stages tends to be given greater emphasis than others. Administrators and organizations adapting to change emphasize the reflective what should be stage, while scientists remain focused on observations of what is. Successful managers of social change projects and holistic thinkers tended to make the imaginative leap to what could be. The professions remain concerned with the traditional outcomes of their craft, with what can be. Thus, fragmentation between the learning stages needs to be remedied in any collective learning enterprise.

The ideals-into-practice learning spiral will need tools firstly to ensure equal contribution from all the interests at every stage, and secondly, to maximize the mutual learning within each of those stages. An A – Z list of possible tools can be found in Part 3 of this Guide.

> **Box 1.4 Stages of Collective Learning**
>
> Stage 1. What should be?
> Ideals: translating different ideals into shared principles for action asks for mutual acceptance of difference.
>
> Stage 2. What is?
> Facts: identifying the supporting and impeding factors for collective learning means acceptance of different points of view.
>
> Stage 3. What could be?
> Ideas: bringing together different creative ideas calls for the groups involved to celebrate their differences.
>
> Stage 4. What can be?
> Actions: combining different contributions to collaborative action creates a whole more effective than any one part.

Handwritten annotation:

6 STEPS TO EFFECTIVE COLLECTIVE LEARNING:
1. SETTING THE SCENE
2. DEVELOPING IDEALS
3. DESCRIBING FACTS
4. DESIGNING IDEALS
5. DOING IN PRACTICE
6. FOLLOWING ON

After fifteen years of trialling the collective learning spiral, the Local Sustainability Project has made the following adaptations of the original individual learning cycle.

Reflection: The collective learning cycle depends on the ability of all the contributing groups to reflect on both their own and the others' learning. Thus each learning stage includes an avenue for reflection, and there is time allotted to reflect between each of the learning stages. The outcome of this is more like a collage in which all the contributions are valued, rather than a pre-determined jigsaw puzzle in which the pieces are made to fit.

Order: In developing collective learning for transformational change it became clear that it is essential to follow the four stages in a definite order. Starting with strongly-felt ideals on what should be is essential in that it ensures that the collective learning is driven by a desire for change and a vision of what that change might be. The next stage, what is, establishes the range of facts that allow for the opportunities for and blocks to change in the light of the ideals, rather than being fixed in the present problems.

For the first and second stages, the interest groups contribute from their own positions, thus enlarging their understanding of each other. In the third stage, what could be, they come together to build on this understanding, creating what David Bohm calls 'learning from difference, not more of the same'. For the final stage of each collective learning cycle, what can be, it is essential that a collaboration is built on all the learning that went before, rather than perpetuating old divisions. The field studies confirmed Kolb's insistence that no learning occurs without this final step.

Culture: The learning styles and stages are primarily based in Western culture. However, to the extent that the experiential learning cycle reflects all human adult learning, the cycle proved to be effective with Indigenous Australian communities and in Indonesia, Malaysia and Hong Kong. Great care needed to be taken to ensure that the non-Western peoples re-designed the cycle in their own terms. Examples of application in other cultures can be found in Case Studies 10 and 16.

Knowledge: Relationships between learning, knowledge and power are so intertwined that a separate discussion is needed to apply them to conditions for transformational change. In Chapter 10 the relationships are reviewed from the perspective of the person or group with the responsibility of guiding the collective learning.

Ground-truthing: In satellite monitoring from space, the phrase 'ground-truthing' refers to checking the effectiveness of the aerial measures on the ground. This is equally necessary in applying the stages of collective learning. The first three stages develop the capacity for a guided transformation. Only putting them into practice at the same time, in the same place with the same people will enable the transformational change to take place, and with it the collective learning that allows the learning cycle to continue.

In this chapter we have explored the ideas behind the individual and the collective learning cycles in relation to transformational change. The emphasis throughout has been on experiential learning, 'learning by doing', and on the importance of welcoming diversity. In Chapters 2 – 9 theory is translated into practice in terms accessible to diverse participants. Some of the same material is repeated in terms that can be used in an actual transformational change program. Chapter 10 returns to theory for the benefit of anyone guiding deliberate transformational change.

2 The practice: Party time

Why this Guidebook?

> Summary: This chapter translates Chapter 1 into practice. It identifies the rules, the framework and the principles for celebrating and guiding transformational change.

In this Guidebook we take up the challenge of guiding transformational change. The complex environmental, social and economic issues of the 21st century can no longer be resolved by repairing existing systems. There have been many inadequate responses to climate change, the obesity epidemic, toxic pollutants, shortage of fresh water and urban violence. Not surprisingly one-off stop-gap solutions are failing to take hold at either the global or the local scale. Collaboration alone cannot resolve these complex problems. In resolving society-wide issues, transformational change based on collective action has become not optional but a necessity, not a matter of avoiding but of celebrating change.

Exploring the potential for change in a complex issue is a very different task from standard problem-solving. It is usual to look for cause-and-effect and expect one right answer. By contrast, we need to treat the capacity for transformational change as a 'wicked problem'. A wicked problem resists attempts to resolve it, is the concern of multiple interests, and cannot be resolved without changes in the society that produced it. The situation matches the proposition often attributed to Einstein 'You cannot find solutions to complex problems from within the thinking that created them'.

Blocks to transformational change are well-established in our Western society. First, in spite of the constant calls for their integration, our institutions still separate environmental, social and economic concerns. Second, collaboration between communities, relevant experts and government has proved a major challenge for our competitive and fragmented society. Third, governance is disconnected between the local, regional and national scales. Examples of two way top-down and bottom-up collaboration are rare.

Deliberate transformational change needs a direction. Throughout this Guide it is assumed that the direction is towards a more just and sustainable future, a journey rather than a pre-determined destination. Guiding the changes requires a process which incorporates diverse interests, multiple sectors, and overlapping scales. Such a process needs to celebrate the future possibilities from combining diverse interests; and guard against the negativity from trying to resolve past conflicts of interests. This shift in focus is a transformational change in itself.

The Guide is presented in three parts. Part 1, Collective learning for transformational change treats transformational change as mutual learning among all those with an interest in an issue. The six steps of the collective learning process allow a project team or a project leader to bring together all the contributing interests and to address the full complexity of a project. Instead of 'meetings, bloody meetings' all the interests come together in a convivial atmosphere in a collective learning cycle (Figures 2.1 and 3.1).

6 Knowledge Cultures

- Individual
- Community
- Specialized
- Organizational
- Holistic
- Collective

Collective Learning Cycle Stages

Ideals → Facts → Ideas → Actions

Collective Learning from A-Z

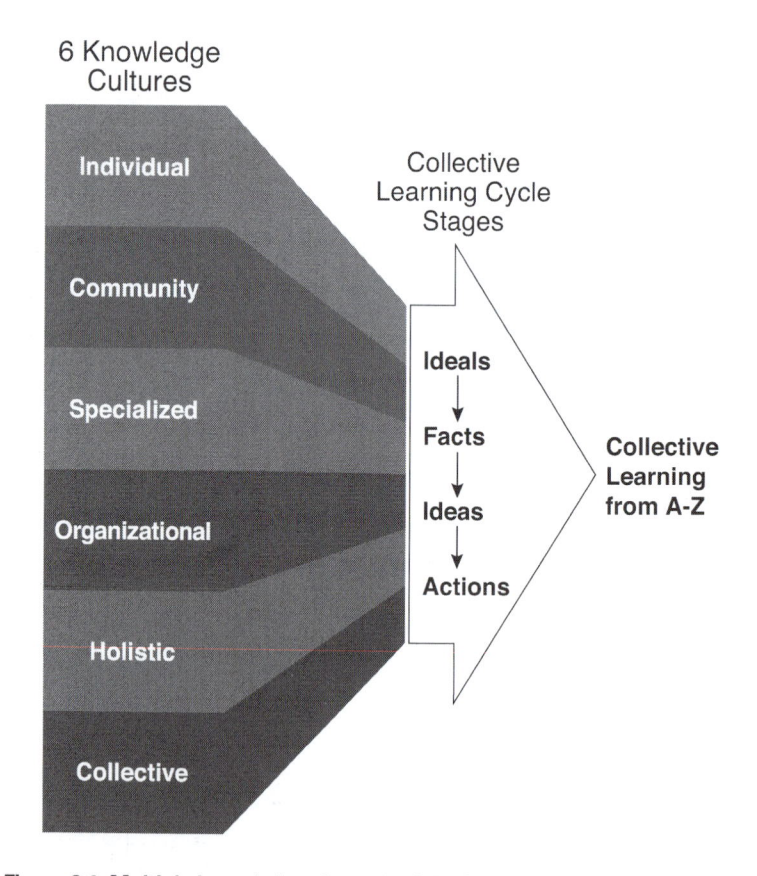

Figure 2.1 Multiple knowledges in each of the four stages of collective learning

Part 2, Celebrations of collective learning presents case studies of 16 ways for treating guided transformational change as if it were a party. The celebration approach overcomes the traditional negativity when diverse interests meet. The case studies cover transformational change in individuals, groups, cities and regions. The types of intervention range from long-term change to a short term intervention. This variety allows a collective learning team to choose the right type of celebration for their particular purpose.

Part 3, A–Z of Collective learning contains 33 integrative tools on which the collective learning team can draw as they move around the learning cycle. They range from Adaptive management and Alliancing to Yarning and Zany ideas.

Who are you and who are we?

To Guidebook users:

Who are you and why are you interested in transformational change?

Do you work on complex social, economic and environmental issues?

- For yourself?
- As part of a community?
- As an expert in your field?
- On behalf of an organization?
- As the integrator of a team?

If the answer to any of these is yes, you are likely to be involved in transformational social change.

As a CEO, facilitator, program leader, policy-maker or individual who is initiating transformational change have you ever:

- **said:** these bloody collaborative meetings are a waste of time?
- **thought:** how to stop this project bogging down in old agendas?
- **felt:** if only the program partners would stop fighting?
- **wondered:** how is it other groups come up with such great ideas?
- **asked yourself:** whatever can I do to make this project work?

If the answer to any of these is yes, you are likely to be looking for guidelines on how to work collectively in a hoped-for transformational change.

From the authors:
Who are we and how are we concerned with transformational change?

We work in collaborative action research (Valerie) and community development (Judy). Over the years we have worked together on whole-of-community change with individuals, neighbourhoods, villages, towns, cities, regions, professions, policy communities, and voluntary groups. We have approached each project as a celebration that brings together diverse interests in a guided transformational change.

We take transformation to mean both the change process itself and the outcome: methods of whole-of-community change that create a turning point between the past and the future.

We have learnt that complex issues affecting the whole community call for transformational change in the community itself. Examples of such issues in this Guide are drawn from the many interactions we have observed in practice.

The aim of collective learning is not to find one right answer, nor is it to reach a consensus: it is to value each contribution. Bringing them all together generates a better solution than any one contribution alone.

Collaboration among multiple interests to bring about transformational change calls for a capacity for collective learning. Each contributor welcomes the opportunity to learn from the others. The process of collective learning is meant to be a celebration, not a chore.

The case studies describe how multiple interests work in a spiral of collective learning cycles to make complex socio-environmental changes. The examples cover contexts from individual change to whole cities and towns. The learning cycle linking multiple knowledges has been used cross-culturally, and so it seems that it can be applied to addressing any complex issue anywhere.

SALLY AND RICHARD 1

Meetings, bloody meetings

Meet Sally and Richard who will guide us through the Guidebook. Sally is administrator of a large government department with millions of dollars to spend on good works; that is, good works so far as her Minister is concerned. If you meet her at work she is highly efficient, a tough cookie, hard to convince of anything that is not already on her agenda. If you meet her socially, she is a barrel of fun, ready to party any time, and very good company.

Richard works in community development, by way of having been a ranger in a national park. He found that he was good with people and interested in initiating changes that might make the world a better place. Now CEO of a small but influential social service organization, he has a loyal band of volunteers, a small but expert staff and lofty ideals.

Sally and Richard must work together to deliver a major project intended to resolve social and environmental issues in struggling communities in a fast-changing region. The trouble is, while Sally and Richard are both good at their jobs, they have different personalities and inhabit different realities. Sally wants to please her Minister, Richard to give birth to a model community.

Each community has ideas of its own and specialist advisors offer competing solutions. Meetings are a matter of everyone pushing their own barrow.

Sally and Richard recognize the need to share their skills to find a new, more productive way of doing things. They want to avoid more difficult and unproductive meetings.

The Practice: Party Time 25

Changing the game

Think Journey Not destination

Source: Art of Moving (2006)

Current change management practice seeks to reduce difference rather than welcome it. This immediately puts the players in competition with one another, thus making the problem worse. A collective solution directed to a transformational change puts change management into reverse. Collective learning welcomes diversity and takes a journey that arrives at a multi-faceted answer that satisfies the needs of all the players. The story of Sally and Richard in 'Meetings, bloody meetings' is seeking such a change.

Three sets of rules transform a formal collaboration into a convivial collective learning spiral. The sets of rules are: one, to maintain a welcoming space for an open-ended discussion; two, to collaborate through mutual dialogue; and three, to celebrate the diversity of the interests drawn into the process.

Note: The rules are presented full page for photocopying.

Rules to get to the learning spiral:
1. *Welcoming space for open-ended discussion.*
2. *Collaborate through mutual dialogue.*
3. *Celebrate the diversity of interests brought into the process.*

Rule 1.
Open Space Technology

In a convivial learning process everyone accepts that

- These are the right people
- This is the right time
- This is the right place
- Everyone commits themselves to the task
- Whatever happens is what was meant to happen
- The law of two feet:
  if anyone is not learning, they should move on.

People may not agree with one another, that is fine, but if anyone feels that they are not learning, they should move on. This leaves the rest to take part in a convivial celebration of their collective learning.

© 2013, *Collective Learning for Transformational Change*, Brown and Lambert, Earthscan from Routledge

Rule 2.
Dialogue

In an open collective learning process everyone agrees to adopt the Rules of Dialogue*:

- Everyone speaks from the heart
- Everyone listens without judgement
- Everyone learns from each other
- The answer is always yes

A collective learning process is not just a matter of listening. In everyday conversation we often pursue our own interests and do not wait to understand the other's position. In dialogue, the participants suspend disbelief and delay judgement. The aim is to truly hear and learn from what the other speakers are saying.

* from David Bohm's book *On Dialogue* (1996)

Rule 3.
Diversity

A collective learning process seeks to maximize not reduce diversity. This means establishing:

- Mutual trust

- Mutual respect

- An inclusive language

- Use of the imagination

- An open mind

We learn from difference, not from more of the same. Working with a diverse range of interests can result in tensions and stagnation; or it can provide a rich pool of ideas from which everyone gains.

© 2013, *Collective Learning for Transformational Change,*
Brown and Lambert, Earthscan from Routledge

SALLY AND RICHARD 2

A leap into new territory

Sally and Richard went for help. First they looked inside their respective organizations. Each of their organizations told them not to give in to the other. Sally and Richard were both experienced enough to recognize where they were heading – down the slippery path of confrontation and stalemate. So they brought in a management consultant. She designed an efficient task-centred management structure.

Sally thought it wasn't strong enough and Richard found it too restrictive. Nevertheless, they started to put the new structure into effect. The entire project bogged down in formal meetings and attention to detail. The various interests became further locked into their fixed positions.

Richard suggests that they try a new approach that brings together Sally's management skills, his community experience, expert advisors and the support of their organizations. Their aim is to facilitate a convivial, collective learning environment. Sally remained unsure but was willing to give it a try.

3 Following the collective learning spiral

Summary: This chapter presents the basic principles of a collective learning cycle.

1. Feeling

DESCRIBE

1. WHAT SHOULD BE
from individual to a collective set of ideals

2. WHAT IS
grounding in reality- helping and hindering factors

Watching 2.

DEVELOP

Focus Question ?

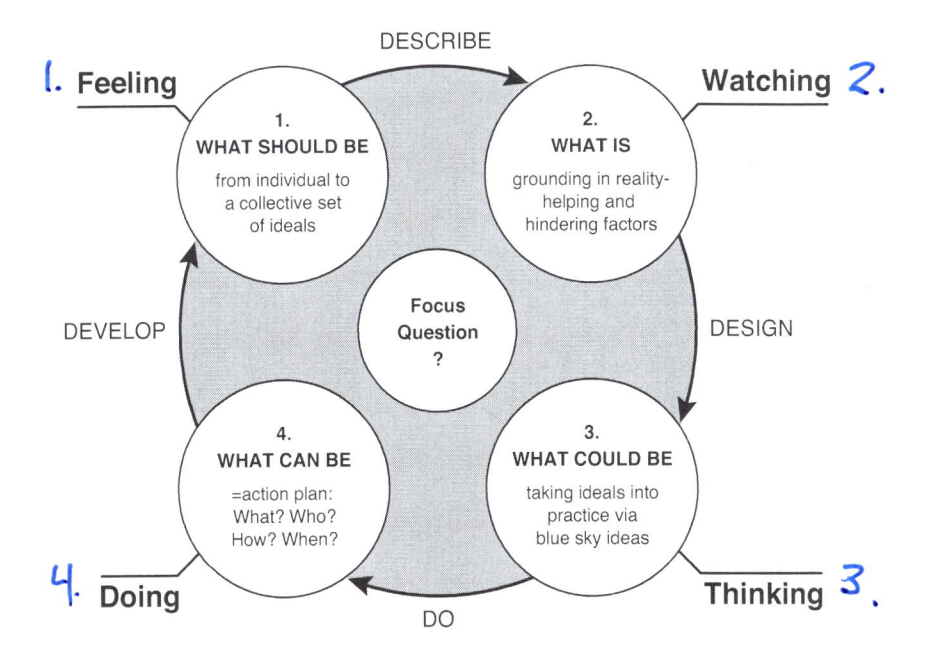

DESIGN

4. WHAT CAN BE
=action plan: What? Who? How? When?

3. WHAT COULD BE
taking ideals into practice via blue sky ideas

4. Doing

DO

Thinking 3.

Figure 3.1 Four Stages of the collective learning cycle = one turn of a spiral

We are about to follow the collective learning spiral as a vehicle for transformational change.

We have generated three sets of rules for setting up each turn of a collective learning spiral. Each of the turns is a collective learning cycle based on decades of work on experiential learning by David Kolb and his associates, and over 300 applications in Australia and beyond (see the case studies in Part 2).

Kolb and others have found that, in practice, adult learning and the related social change takes place in six steps. If each learning cycle is not completed, no permanent change takes place (Figure 3.1). Since people learn from experience, each run of the cycle will carry the learning further. The cycle becomes a spiral in which the learning never ends (Figure 3.2).

The adult learning cycle follows the sequence of setting the scene, clarifying existing ideals, establishing the facts, generating new ideas, taking collective action and following on. The learning cycle turns into a learning spiral when the participants in the initiative build on their collective learning, repeating the cycle over weeks or months or even years.

Collective learning is a framework, not a recipe, since each set of participants makes their own contributions to each of the learning stages. Nor is it a one-off event. It is a continuing celebration which generates excitement and a commitment to change. For collective learning there needs to be agreement on a focus question, the completion of the full cycle through Stages 1–4 below, and follow-up support. At each step fun activities allow individuals to enjoy each other's company.

If a simple collaboration is your goal, then you will find that each learning stage can be used on its own. Stage 1 can be used for clarifying and sharing values, Stage 2 in a broad scoping of the context, Stage 3 in tapping into group creativity, and Stage 4 in the design of a collaborative action plan.

For collective learning all six steps, including all the four learning stages must be completed for each learning cycle. All the interests in the issue are not only included in but enjoy sharing every stage. After this experience participants are ready to continue to work together over the long term.

Handwritten note:

FOR COLLECTIVE LEARNING TO OCCUR, THERE MUST BE:

1. AGREEMENT ON A FOCUS QUESTION.
2. COMPLETION OF ALL 4 STAGES OF THE CYCLE.
3. FOLLOW-UP SUPPORT.

The six steps of each collective learning cycle

Step 1
Setting the scene

The four learning stages and the links between them are described in detail in the sections that follow this one. Here we will go round the cycle once and confirm that it follows plain common sense.

First the participants are asked to agree on a core question that includes all their interests. This might be 'How can we design this learning cycle so we can all enjoy learning from each other?' or it may be a shared question that relates more specifically to the issue that has brought them all together.

At this time, the participants don't have to agree on the problem or the solution, only that there is an issue that involves them all. It is often a relief just to acknowledge the problem even when they almost certainly don't agree on a solution.

Having set the stage, the learning stages begin.

Step 2
Learning stage 1: Feeling

Stage 1
WHAT SHOULD BE
from individual to a collective set of ideals

Clarifying aims.

The cycle starts with participants sharing their ideals for the program: what should be?

The session involves learning about each other's underlying hopes in taking part in the initiative. Answers are expressed as aims, purpose, or ideals. The session is managed so that the aims of every participant in the process are given equal time and equal respect. None are 'right' or 'wrong'.

Step 3
Learning stage 2: Watching

Stage 2
WHAT IS
grounding in reality- helping and hindering factors

Describing the facts.

Participants establish an agreed pool of facts that set the parameters for achieving the ideals: what is?

Each participant contributes what they regard as the facts of the matter from their own experience

Following the Collective Learning Spiral **33**

of the shared issue. There can be no debate, only clarification, since each contributor describes the parameters of the change in relation to their own particular purpose. It is important that both positive and negative factors are included. There can be a sense of relief when the facts are on the table, no matter how depressing.

Stage 3
WHAT COULD BE
taking ideals into practice via blue sky ideas

Step 4
Learning stage 3: Thinking

Generating ideas.
Participants brainstorm innovative ways to achieve their ideals while taking due account of the facts: what could be?

This is the point where blue sky ideas are welcome. Rather than continue to think separately about the issue, here everyone joins in and contributes their own experience and skills. Answers contain fresh perspectives created through dialogue between members. There is considerable excitement in generating new ideas.

Stage 4
WHAT CAN BE
action plan:
What? Who?
How? When?

Step 5
Learning stage 4: Doing

Taking action.
At this stage the whole group develops a collaborative action plan for collective action: what can be?

Checking the new ideas against the ideals and the facts will show which of the new ideas are relevant and practical. Those that survive that test will move on to be the subject of an action plan. The participants design the actions that they are willing to undertake together. Answers are in the form of specific action plans, stating objectives, people, resources and a timetable. If the process of arriving at the action plan has been a positive experience, a working group will be ready to support each other in a mutual project.

Step 6
Following on

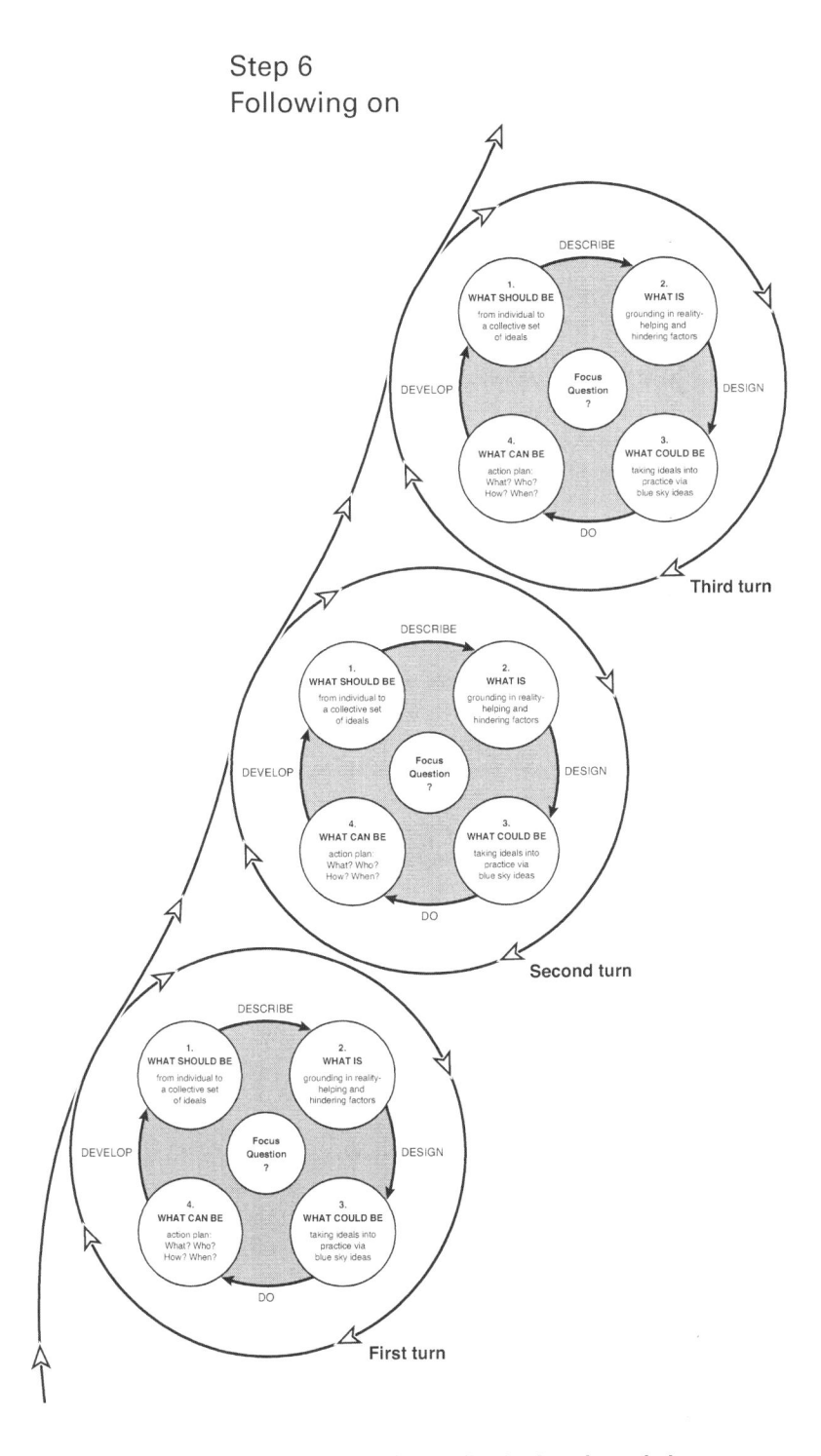

Figure 3.2 Three turns of the collective learning spiral

Following the action, things will have changed. At least, we hope so. This can lead to considerable tension between the new direction, and the pre-existing state. This is like putting new wine into old bottles – the result can turn the new wine quite sour!

The change needs to be supported for some time after re-entry into the conditions that led to the change. The support need not be by the original team but may come from the resources gathered by the new initiative. The details of this stage are described in each of the 16 case studies. The follow-on can be a continuum of programs spawned from the original cycle. Or it can be a training program for new teams who introduce cycles of their own. Or the original team can run the cycle again, three weeks or three months later depending on what the program needs. The spiral continues, changing the ideals in the light of new knowledge gained from the action (what should be, now?).

For practical purposes, community educators refer to the learning stages as:
1. Starting where the learner is at
2. Making a reality check
3. Introducing new ideas
4. Applying the new in the context of the old.

Timetable

The most usual timetable for a transformational change project is three months: six weeks' lead up time (where to begin), a one-day workshop and six weeks' follow up (where to go next). The most effective have three months' lead-up time, a two-day meeting, and six months' follow up, repeated at yearly intervals and/or spreading out through the community itself.

The whole transformational change program may run over three days, three months or three years, according to the program design. Case studies in Part 2 include examples of each of these. In every program or project, however, the time allocated for moving round the cycle should include a lead up, a workshop and a follow on. The workshop session should be divided equally among the four learning stages. For example, in a three-hour workshop, there would be 90 minutes for each stage. In a three year program there may be one learning cycle a year, each with a two day meeting. Each stage would then be allotted half a day.

Applying the rules

The following principles and practices apply throughout the collective learning process

Principle 1 Everybody treats each other's knowledge equally. The process involves a collective of individual people each with different 'knowledges'. Each person's ideas are to be respected and included.

Principle 2 A common goal binds the collective. Every participant in a collective should think of themselves as a valued part of a collective working towards a shared and common purpose.

Practical suggestion 1 The way to keep the collective together is to determine the shared goal at the start and stick to it throughout the process.

Practical suggestion 2 The process should involve time for self-reflection and mutual sharing at every stage. These can range from formal exercises to 'yarning', that is each telling stories about their own experience.

Where is the 'real' reality?

Sally is a realist. She assumes that 'reality' exists and that the task is careful management of the problem-solving process. Richard is a relativist who believes that there can be many interpretations of reality, and participants in a change process will each construct their own version.

We are assuming here that they each have a point. Collective learning needs both firm organization and openness to new ideas. Throughout this book we assume that there is a 'real' reality outside ourselves. However hard we try we can only partly know it, although we do the best we can.

The task in collective learning is to have the full set of contributors to a decision create a synergy towards their common goal. The next chapter explores just what we mean by 'a full set of interests' in a decision.

SALLY AND RICHARD 3

My goodness! It worked!

Richard and Sally trialled the collective learning process in a pilot project for their main program. The project was to develop a community-based health centre in an outer suburb. The project involved leading community supporters, local community organizations, health experts and the sponsoring organization. Each had their own vision of what the centre would be like.

All invitations were accepted, and the focus question became

> *How can we have a health centre which meets the needs of professional staff, its sponsoring organization and its community?*

In stating their aims for the project, community members wanted 'safe births, easy family access to local practitioners, and convivial communication with staff'. A medical practitioner wished for 'hygienic conditions, efficient administration and support staff trained in Western science'. For the sponsoring organization, the government aim was 'financial accountability, operational improvement, and capacity building'. These very different agendas produced very different sets of facts and ideas. This put considerable pressure on Richard as facilitator and Sally as content supervisor to keep the process moving forward.

After moving through the learning stages, the group designed a multi-faceted action plan which called for a supportive team, something that would continue after the health centre was in operation.

Sally recognized the importance of good facilitation in creating energetic participant sharing of ideals and the enthusiasm with which action plans were developed by the group. Richard saw Sally's organizational skills and attention to the focus question as a key to the success of the pilot project.

4 Step 1
Setting the scene: Who to invite?

Summary: In this chapter, the key issue of who makes up the collective is resolved as the full set of decision-makers in a transformational change: key individuals, the affected community, relevant specialists, influential organizations and creative thinkers.

Who to invite?

One of the first questions to ask when starting out on a collective learning process towards a desired transformation is 'What is the issue?' and then 'Who do we invite in order to achieve the change?' There will usually already be a high profile issue as the reason for meeting. It may be a local issue: building a new park; it may be the effects of a global issue: storm surge from climate change; or it may be the global issue itself: a sustainable development program for a region. In any case, for collective learning the diverse participants need only to agree that there IS an issue.

Since the participants will bring their own interests with them they will bring their own version of the issue to the table. Therefore the program team can't define the issue at the start of the process, only agree on a shared question. So having agreed that there is an issue, the question for the first step of the collective learning spiral is whom we should invite? Do we invite the whole community, the leading players or a selected group? The answer is that it can be all or any of the above, according to the scope of the issue. In every case, however, the important question is rather, have we invited people representing all those who should be involved in making decisions about the issue?

Figure 4.1 The parties involved in making any lasting transformational change. Research in over 300 communities confirms that these are:

Key individuals

Affected communities

Specialist advisors

Influential organizations

Holistic thinkers

The ideal is for a for the full set of 'decision-makers' to be present. This diversity ensures that all the partners able to effect the change are present. Sometimes this is not possible, so choices have to be made. How many of the interest groups will accept the invitation? Can some of the participants role play the part of the missing ones?

Each of the key set of 'decision-makers' has their own goals, accepts certain types of evidence, has existing knowledge content, and its own language. In other words, we find that we are dealing with different knowledge cultures, each with its own interests (Figure 4.1)

Collective Learning for Transformational Change

Left to themselves, each of these knowledge territories has become sufficiently self-contained to develop as a different culture. In any meeting, the separate interest groups are more likely to form a clique than to collaborate in a collective understanding (Figure 4.2). Throughout a collective learning process, the rules of open space, dialogue and diversity ensure that each of these knowledges is able to hear each other and to contribute equally.

The knowledge cultures of western decision-making (Brown 2008)

Knowledge Culture	Structure	Sources of truth	Sources of ignorance
Individual Knowledge Lived experience, identity		Memory Learning style Five senses	Subjective Limited Vague
Local Knowledge Shared experience of people and place		Stories Events Symbols	Gossip Anecdote Inaccurate
Specialized Knowledge Mono, multi and trans-disciplinarity, the professions		Inquiry Measurements Observations	Jargon Irrelevant Narrow
Organizational Knowledge Administration, government, industry, strategic thinking		Agendas Alliances Networks	Deals Mates Corruption
Holistic Knowledge Essence, core, purpose		Synthesis Focus Creative leap	Airy-fairy Impossible Impractical

Figure 4.2 The knowledges needed for transformational change

Figure 4.2 uses a set of symbols to describe each knowledge culture or interest group. Individuals are represented by scattered dots, indicating their wide spread of interests. Community is indicated by a wavy line, suggesting that while communities are diverse, they are linked by core characteristics found in each functioning society. Specializations are represented by a string of unconnected boxes, showing the individual frameworks of the different disciplines. Organizations are closed circles moving in one direction, indicating that each organization already has a strategic direction. In the centre is a star, standing for a holistic understanding of the project that includes them all (Figure 4.4).

Figure 4.3 Individual positions on transformational change: community member, expert, politician and holist

However, each knowledge culture has a habit of rejecting the others' forms of knowledge. Individual knowledge is called only anecdote, community knowledge just a story, expert knowledge fragmented jargon, organizational knowledge self-serving and holistic knowledge too airy-fairy. This approach to diversity creates considerable challenges, as represented in the shouting figures in Figure 4.3. Part 2 gives 16 examples of the different ways in which those challenges have been met.

The number involved in any collective learning cycle depends on the aims of the program/project. What is important is the range of interests being present. The process has been run with 7, 15, 30, and 200 participants.

You will have noticed that we are not using the term 'stakeholders' for the different interest groups. Stakeholder implies an existing interest in the program focus question. In a transformational change there are interests that are often excluded. Examples are radical thinkers, children not yet born, and activists protesting against the current solutions. More important than involving only the well-known stakeholders is to include the full set of those affected by any significant change. Figure 4.4 is a mandala for collective learning: reflecting a process in which each knowledge culture builds on the others. This is not a hierarchy – every layer is of equal importance.

42   *Collective Learning for Transformational Change*

Knowledge cultures as a nested system

Culture and content	Symbol	Content
Individual Knowledge Own lived experience, lifestyle choices, learning style, identity		Reflections Learning
Local Knowledge Shared lived experience of individuals, families, businesses, communities		Stories Events Histories
Specialized Knowledge Environment and Health Sciences, Engineering, Law, Philosophy, etc		Case studies Experiments
Organizational Knowledge Organizational governance, policy, strategies		Agendas Alliances Plans
Holistic Knowledge Core of the matter, vision of the future, a common purpose		Symbol Vision Ideal
Collective Knowledge		

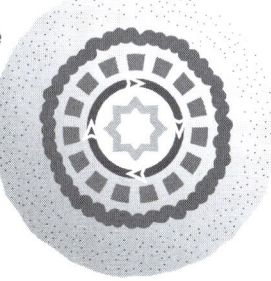

Figure 4.4 The mandala of collective learning

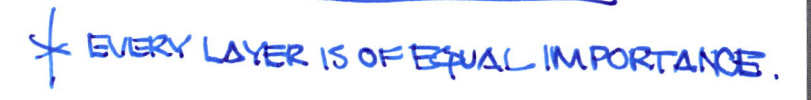

EVERY LAYER IS OF EQUAL IMPORTANCE.

The Tools

An extensive portfolio of tools helps choose how and when to work with the knowledge cultures so that they are in a collaborative mood. Some of these are described for each stage of the collective learning in Chapters 4–9. Sixteen case studies give further examples of collective learning in practice in Chapters 12–19 of Part 2. A set of tools useful for every step of the collective learning cycle makes up the A–Z of Collective Learning Part 3. The leader or the group choose the right tool for the right step of a cycle.

Six steps, four learning stages

Step 1
The invitations, the focus question and the tools

The first step in designing the learning process has been to decide on the issue and the guest list. The next step is to bring the interests into a shared learning environment as soon as possible. Before you design the invitations put together a pilot group with members of all the interest groups. This group can advise you of whether your language and exercises are compatible with their respective knowledge domains.

Their first task is to develop a project focus question that will appeal to all participants. Since this will be the driver for the collective learning cycle, it is important to get it right. We have already suggested two possible focus questions and examples can be found in each of the case studies.

- A great deal of progress could be made before people even meet. Using the social media, the participants can be introduced to one another and asked for their individual goals for the project. They can also be asked to set the agenda.

Once you have the focus question you can design the invitation, again in a language open to each interest group. An invitation in technical terms may seem very clever, but is unlikely to attract community members. Policy language is well-known for putting everyone else off.

An example of this put-off is a local government brief to discover the preferred futures of a string of coastal villages in their jurisdiction. The invitation to discuss the future of 'Southern Beachhaven' as in the brief fell on deaf ears. Further inquiry brought the reply 'Southern Beachhaven – there is no such place. Only the Council would ever call us that'. It turned out that the separate villages were fiercely proud of their individual identities. Once that was taken into account in our language and our communications, the project went swimmingly.

Messages are transmitted in ways other than speech. The environment in which the project participants meet conveys a lot to them about the project. A luxury conference hotel may be just right for one project, and a pretentious mistake for another. Whatever is chosen, there needs to be a sense of valuing the participants and seeing to their comfort. Refreshments on arrival and good food at all meetings ring positive bells in every evaluation.

Whatever the setting, there should be room for round tables seating 7–10 people, with enough space for them to both hear each other and be close enough to see and hear a central speaker's table. The documents described in Chapter 5 under the four learning stages should be ready on the table, with a group membership list, the learning diagram, and the aims and timetable for the meeting. Water and peppermints seem also to be obligatory!

There are four roles vital to the success of each workshop:

- A Chair who welcomes people, introduces the project, backs up the workshop team and praises the action plans. This doesn't have to be a grand guru, just someone who is acceptable to the different interest groups and knows what the job is.

- A Knowledge-broker who presents the ideas behind the meeting, follows the ideas as they develop, constructs the summaries of each learning stage, and introduces fresh ideas and activities as needed.

- A Facilitator with people and negotiating skills who explains the process that the workshop will follow, ensures that each table is proceeding smoothly through the exercises and that all voices are being heard.

- An Administrator who sees to the organization of the workshop, the publicity, the mail outs, the lists of participants, the needs of the participants, the briefing papers and the food.

In small projects these roles may have to be taken by two people or even one. Usually there are friends of the project prepared to take them on. The roles are better separated, since trying to arrange for refreshments to be on time and taxis booked, keeping the groups working smoothly, ensuring that the ideas stay on track, and acting as the up-front host is a difficult task for any one person.

A useful tip is to have some reward or ongoing resources ready for the Chair to announce at the beginning and again at the close of the process. As people prepare their action plans they commit to considerable further work, so it is a good idea for them to know that their work will be rewarded in some way.

Step 1: Setting the Scene: Who to Invite? **45**

Grounding the process: choosing the tools

While the overall learning process has been thoroughly tested and confirmed as a reliable structure, the focus question, the setting and the participants are highly individual matters, redesigned for each project.

Each step of the learning cycle will need the choice of the right tools to help each particular group work on their own issue. Useful tools can be found in all three sections of this book.

The tools found to be most generally useful are in Part 1, linked to each step of the learning cycle. Specific tools used in particular contexts are described in the individual case studies in Part 2. A wide range of tools that can be used at any stage, wherever appropriate, make up Part 3.

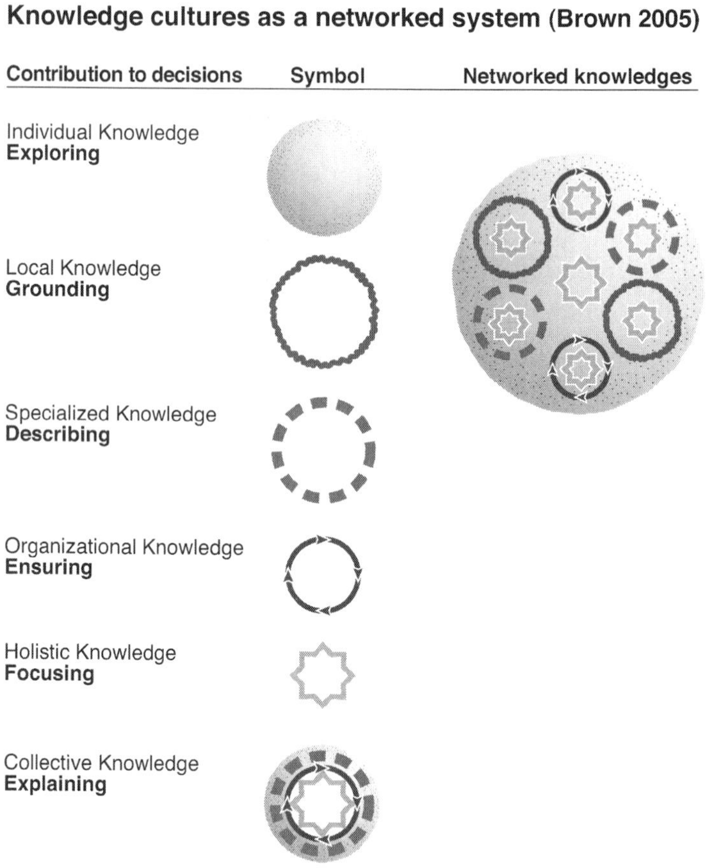

Knowledge cultures as a networked system (Brown 2005)

Contribution to decisions	Symbol	Networked knowledges
Individual Knowledge **Exploring**		
Local Knowledge **Grounding**		
Specialized Knowledge **Describing**		
Organizational Knowledge **Ensuring**		
Holistic Knowledge **Focusing**		
Collective Knowledge **Explaining**		

Figure 4.5 Knowledge cultures' contributions to a collective decision

The learning cycle is not a cure-all. It fulfils some purposes and not others. The collective learning process is useful for: a. individuals wishing to work on the whole of an issue, b. diverse interests coming together to work on a shared concern, and c. projects that aim to encourage transformational change. It is also useful for building a transdisciplinary team, and resolving a 'wicked' problem, one that requires changes in the society that caused it. The diverse interests each contribute their own skills and ways of working (Figures 4.5, 4.6).

The collective learning process is not useful for: a). long established groups who wish to continue as before, b). policy development or management structure where there is no wriggle-room for change, c). problems where the interest groups are so divided they are not even willing to be brought together for a common purpose, d). cases where other learning processes have been tried and failed and led to extreme resistance, and e). the problem is simple and has an obvious solution.

Figure 4.6 Knowledge cultures can hear each other

SALLY AND RICHARD 4

Now for the big one

With a successful pilot project behind them, Sally and Richard were ready to embark on a major project designed to change the way the whole region thinks about and acts for a sustainable future.

But it was not without some trepidation that they moved to this major task.

After some brief discussion, they recognized the importance of knowing what people in the region saw as the key issues so that invitations to participate were relevant and appropriate.

[handwritten annotation: DEVELOPING THE QUESTION:]

Using the internet and email communications, they identified a small group from all the interested sectors.

Sally and Richard then held a telephone conference with a representative from each of the interest groups involved in making decisions for the region (7 people in all) and discussed the people and the issues. Between them all they developed a focus question that covered all the issues without making any assumptions about the solutions.

The focus question was

> *How can we help our region survive under these conditions?*

The 150 people invited were asked to RSVP and send their particular interest in the topic by email.

With this information to hand, Sally and Richard were almost ready to begin a collective learning cycle, but first they needed their project team to become more familiar with the local region. A briefing from their small advisory group soon provided valuable input and the process could begin in earnest.

5 Step 2
Collective ideals: What should be?

Summary: This chapter describes the opening session of a learning cycle and presents the case for starting the learning cycle with ideals rather than facts.

DESCRIBE

1. WHAT SHOULD BE
from individual to a collective set of ideals

2. WHAT IS
grounding in reality- helping and hindering factors

DEVELOP

Focus Question ?

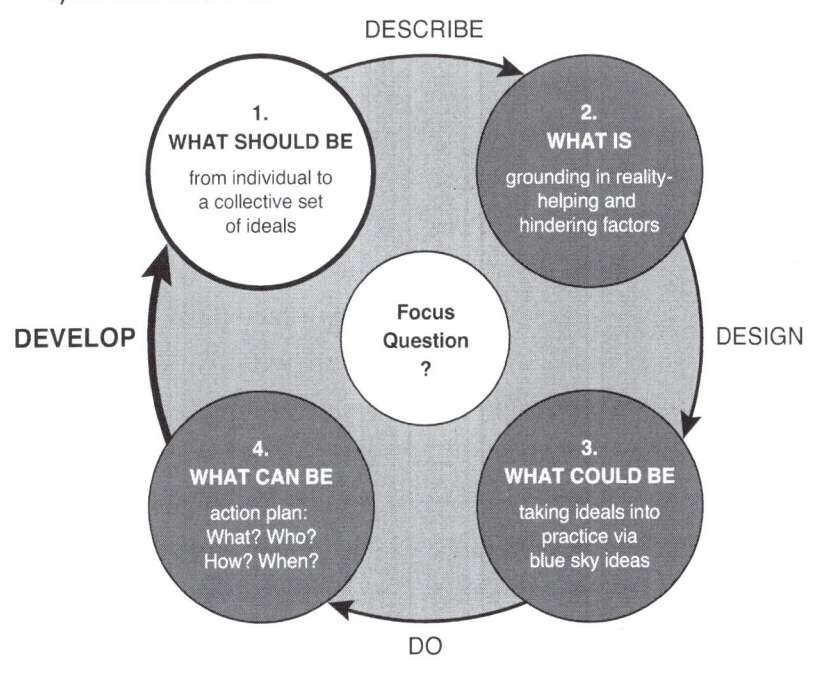

DESIGN

4. WHAT CAN BE
action plan: What? Who? How? When?

3. WHAT COULD BE
taking ideals into practice via blue sky ideas

DO

Figure LC-1 Learning cycle stage 1: developing ideals

As you start the collective learning cycle proper, you will already have arranged the setting. This will be in a pleasant room in an interesting environment with a sense of celebration and excitement at meeting together; or online after some getting-to-know you time. It will be assumed here that the meeting is face-to-face, although the learning cycle has been used online in tertiary courses, and distance learning.

The participants will be sitting at round tables in mixed groups from the five different interests. Each participant will have a name tag or a table tag, and a coloured dot or number to identify their group. This allows the participants to recognize each other and the facilitator to call on each group in feedback sessions. Documents already on the table will include a list of the group's names, a diagram of the learning cycle, and a timetable.

The workshop chair will welcome participants, wish them luck, introduce the change management team, and ideally point out how the outcomes of the collective learning will be put into action. This introduction should be brief (ten minutes) and make it clear that the workshop belongs to the participants, not the host organization.

Participants have been asked to go straight to their table as they arrive. People are invited to swap tables if they wish, although it is up to the facilitator to make sure that there is still the full set of interests on each table. The first step after the formal welcome is always to ask table members to introduce themselves to each other by name and by personal interest in being there. It is important that each person identifies themselves as an individual and NOT as a representative of their interest group. This keeps the discussions at an experiential level and helps reduce old tensions. Responses have ranged from 'to change the world' to 'getting a day off work'. Whatever they are, the responses are always accepted as valid.

The knowledge-broker or facilitator then describes the thinking behind collective learning for this particular workshop, and repeats the potential focus question. Participants are asked to discuss or re-create the focus question, so that everybody feels that they own it. The timetable is reviewed and then everyone settles down to the first question: what should be?

Beginning the process with the participants' ideals already breaks with tradition in two ways.

- **One**, it is more usual to start with 'the facts'. Starting with the facts freezes the issues in time, and greatly reduces the drive for change. Starting with participants' aims and ideals is therefore essential, although this can be the subject of considerable disquiet. 'But we always start with the facts' is a not uncommon objection. Explaining the reasons why is usually enough to get the show on the road.

- **Two**, standard practice is for the sponsoring organization and/or the program leader to tell the participants what the aims are for the program/project. While both of these have their own ideals, they are not alone. The participants also have their own ideals. Not sharing and respecting these at the beginning leaves a tension and a distancing which can sabotage the project.

The marked differences between the aims of individuals, community members, specialists, organizations and creative thinkers (Figure 4.2), are to be celebrated, as we discussed in Chapter 2. However, they all use the same learning spiral. We found in Chapter 1 that this cycle is the basis of personal learning, community development, a research design, a strategic plan, and a creative leap, so all the interests should feel comfortable with a familiar practice.

By starting with sharing their individual aims, the participants are already stepping away from expecting a single preset position on the problem. When each member's ideals are equally respected, discussing 'what should be' raises the likelihood that the collective will end up with innovative and even transformative solutions. And everyone enjoys hearing their ideals acknowledged.

Learning styles. Of Kolb's four learning styles that can be expected to contribute to collective learning (see Chapter 1), it is accommodation that is most likely to dominate the first learning stage. Ideals are based on a combination of feeling and action. Accommodators are attracted to new challenges and experiences, and to carrying out plans. They commonly act on 'gut' instinct rather than logical analysis. This stage draws on Gardner's emotional intelligence, and on Buber's 'I–thou' relationship with people and with the planet (Chapter 10).

Workshop design. The workshop needs to take steps to supply support for participants to access their deepest feelings. The desired outcome is a vision for the future, not a wish list based on what the participants do not have now. Care needs to be taken that convergers do not change the request for ideals into concrete objectives, or assimilators into a tight logical response based in what is now.

Tools. Ask each individual to write down their own five aims for the fulfilment of the focus question. Then share their ideal solutions with the group. They might do this by sharing stories or fill in a set of cards. Several options can be found among the case studies (Part 2) and list of tools (Part 3). Questions are encouraged on points of clarification. Agreement or disagreement is not allowed. When everybody has heard each other's ideals, the facilitator can help the group to organize the sets of ideals into categories and to give the categories a name. Thus the group has moved from a set of individual ideals to a collective set that everyone understands.

Also see: Forecasting, Visioning, Balancing in Part 3.

Step 2: Collective Ideals: What Should Be?

Making the connections between what should be and what is

During the cycle, the arrows that are making the connections between the stages, are as important as the stages themselves. At the close of the 'What should be?' stage, there needs to be a way of summarizing and sharing the pattern that each group has formed to connect their aims. This can be spoken, on butcher's paper so all can see, or thrown up on a screen. Once more, the material is there for explanation and clarification, but not for approval. There can be no right or wrong if all aims are to be respected.

As you move on to the next stage, expect a shift in mood. Be prepared for some strong discussion at this stage. If the scene has been properly set, the diverse participants are usually ready to accept each other's diverse goals. They are much less likely to accept that there are different sets of facts contributed from each set of interest groups.

Be sure to get everyone out of their seats as part of the sharing. This makes for a body language shift as well as a mental shift from values to facts.

Debriefing and reporting

It is useful to test the cycle with colleagues or familiar clients before embarking on a major program, and for the change management team to debrief after each cycle. Further, the learning guides will always learn from the participants as they pass through each stage. If possible there needs to be a debriefing among the team members after each stage. An example of a team debrief can be found in Chapter 14 (Case Study 5).

The whole process will always involve a report to participants and so making notes will help this process. The report can always be easily prepared straight from the group reports at each learning stage. It is desirable, although not always possible, to present the report to participants for validation the next day, or certainly within the week, as part of the transparency of the exercise.

Participants should be encouraged to make personal notes on their own learning at the end of each stage.

6 Step 3
Collective facts: What is?

Summary: This chapter offers ways to collect from all the key interest groups the helping and hindering factors for the transformational change.

DESCRIBE

1. WHAT SHOULD BE — from individual to a collective set of ideals

2. WHAT IS — grounding in reality- helping and hindering factors

Focus Question ?

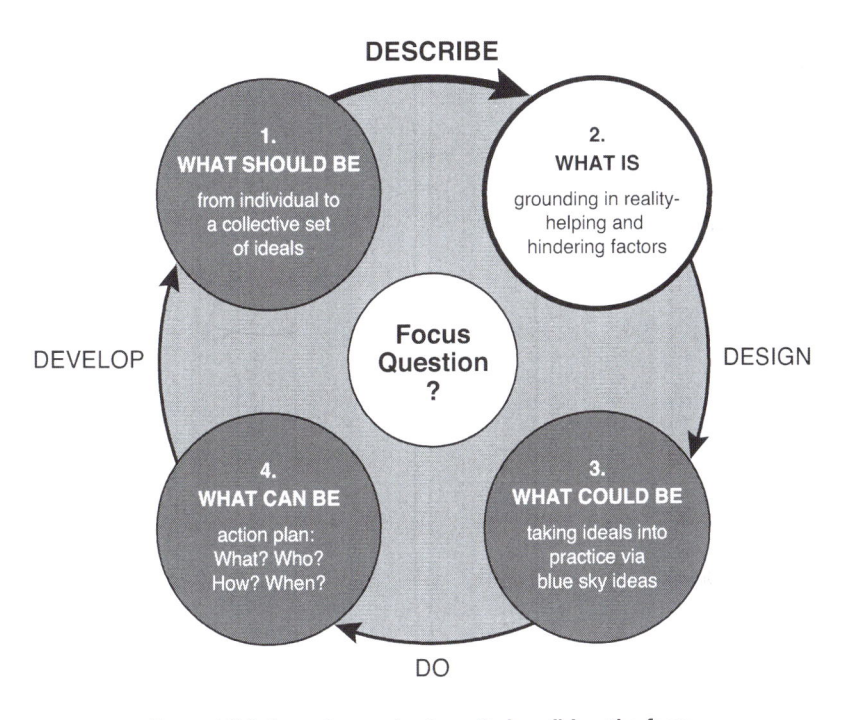

DEVELOP — DESIGN

4. WHAT CAN BE — action plan: What? Who? How? When?

3. WHAT COULD BE — taking ideals into practice via blue sky ideas

DO

Figure LC-2 Learning cycle stage 2: describing the facts

Once the group has come up with a set of ideals in Stage 1, the ideals need to be grounded in reality. Everyone is asked to draw on their own knowledge base to describe what they consider to be 'the facts' that would help answer the focus question.

The facts are separated into factors that help and factors that hinder the achievement of the set of ideals. There is no prioritizing or ranking among the different approaches to 'facts'.

Be prepared for each of the knowledge cultures in the decision-making to draw on a different source of evidence (Figure 4.2):

1. • Individuals draw on their own experience.
2. • Communities draw on shared past events and expectations of the future.
3. • Specialized knowledge is built on objective inquiry and measurement.
4. • Organizations consider compatibility with their own strategic planning processes and agendas.
5. • Holistic knowledge is validated by accepting the creative interpretation of 'the facts'.

Dialogue is needed to ensure that all key factors have been included, and that everyone has contributed.

Tools: A field force map helps everyone find the relationships between the different sets of helping and hindering factors (Figure 6.1).

Hindering factors
(e.g. lack of resources, inefficient organization)

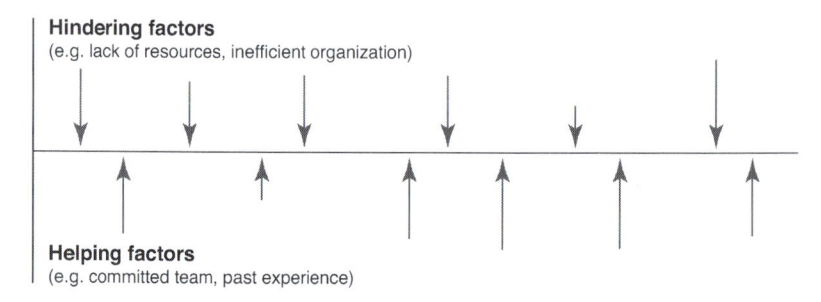

Helping factors
(e.g. committed team, past experience)

Figure 6.1 Field force map

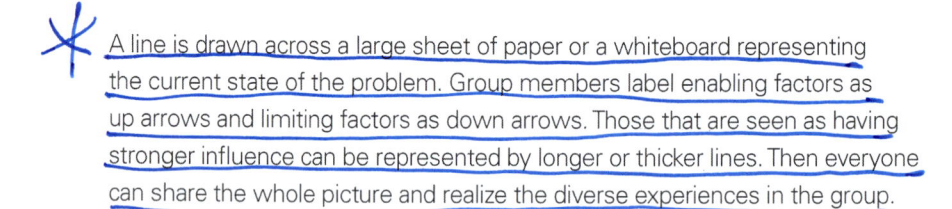

A line is drawn across a large sheet of paper or a whiteboard representing the current state of the problem. Group members label enabling factors as up arrows and limiting factors as down arrows. Those that are seen as having stronger influence can be represented by longer or thicker lines. Then everyone can share the whole picture and realize the diverse experiences in the group.

Since this is the learning stage likely to find disagreement between participants as to what are the facts, see Conflict Resolution and Problem-Solving in Part 3.

54 Collective Learning for Transformational Change

Making the connections from what is to what could be

At the close of Stage 2 the group as a whole reviews the map which sets out the context for change. This allows the group members to share the levels of difficulty and the availability of resources. If there is more than one group, each group should do their own map. Then groups can rotate around each table, and examine the different descriptions of the context of the issue. Don't be surprised if each group context is different.

The shift between Stage 2 and Stage 3 is perhaps the most intensive in the collective learning cycle. Here the group moves from what they already know to explore new possibilities, the seeds of a transformation. The move from facts to new ideas could be described as moving from left brain dominance to more strongly engaging the right brain.

Learning styles (page 9 and Figure 1.2):
What to expect in Stage 2, Step 3

- **diverging (feeling and watching)** - These people in the group will include interesting items that others in the group may not agree are facts. The facilitator may need to support them because these may be important facts.

- **assimilating (watching and thinking)** - People with this style are more attracted to logically sound theories than approaches based on practical value. They may argue strongly that they have superior facts.

- **converging (doing and thinking)** - People with a converging learning style are best at finding practical uses for ideas and theories. Their facts will be drawn from experience and may involve events within the group's experience.

- **accommodating (doing and feeling)** - These people link their facts to personal experience, an experiential approach. They commonly contribute from 'gut' instinct rather than logical analysis.

Tools:

The shift is marked by some creative intervention in the process, which helps the group move from individuals to a team. Case studies in Part 2 describe a range of events, chosen according to the mix of the group and the type of focus question. These have included an artistic installation, a drumming, a sharing circle, and a physical exercise.

A simple exercise allows the group members to know each other better and opens them up to new ideas. Everyone in turn answers the question 'Would you tell us something about yourself that would surprise us all?'. Responses have included 'I was on the Berlin Wall the night it fell'; 'I used to be a powder monkey' (i.e. arm explosives on a construction site); and 'I spent a day with the Dalai Lama'.

Also see: Xing the minefield, Risk and risk-taking in Part 3.

7 Step 4
Collective ideas: What could be?

Summary: In this chapter, participants are asked to move from checking the facts to generating new ideas.

DESCRIBE

1. WHAT SHOULD BE
from individual to a collective set of ideals

2. WHAT IS
grounding in reality- helping and hindering factors

DEVELOP — Focus Question ? — DESIGN

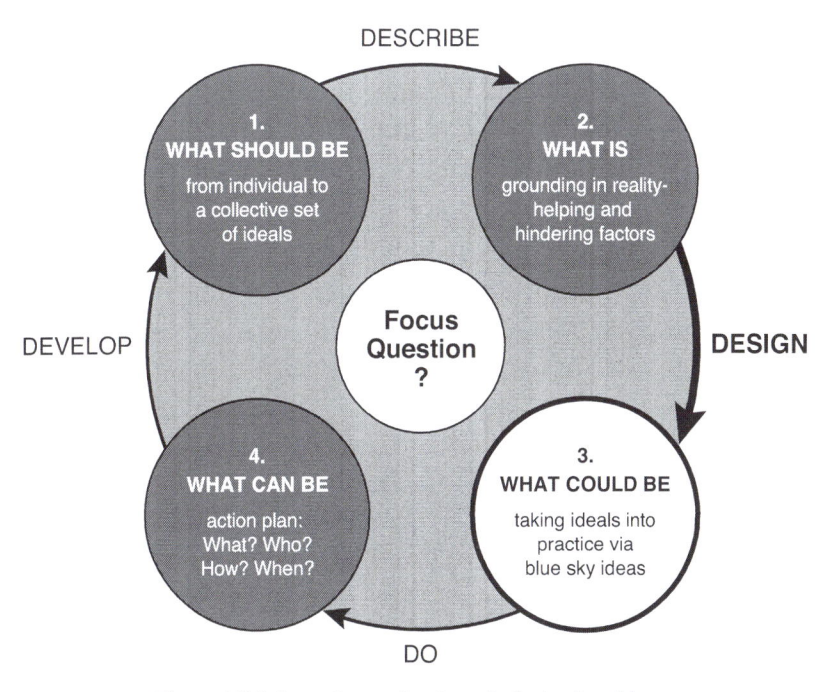

4. WHAT CAN BE
action plan: What? Who? How? When?

3. WHAT COULD BE
taking ideals into practice via blue sky ideas

DO

Figure LC-3 Learning cycle stage 3: designing ideas

Here the facilitator helps the group push the boundaries of their thinking by challenging each participant to come up with fresh ideas for 'what could be' in answer to the focus question. The ideas are grounded in, but not limited by, what is. The conditions for this stage are those that best enable creative thinking: trust, security and challenge. Trust and security are established during Stages 1 and 2. Stage 3 presents the challenge of the use of the imagination.

> **Box 7.1. The Collective Learning Buzz**
>
> Here is an excerpt from a meeting of a Transition Towns group. Transition Towns are a worldwide network of towns searching collectively for creative ways to move to a sustainable lifestyle.
>
> 'It was the fact that when we met up as a group in these public spaces something happened between us. Something we held in common. We understood implicitly what we were doing and why – sharing stuff, organizing events, going through the agenda. When I looked at this working-together in the visioning it looked like an energy field, the kind of energy field you sense when you stand by a hive humming with bees. A hum of warmth and intelligence that allows people to naturally collaborate and make that low-energy downshift happen. When that's going on you don't need possessions to compensate for your isolation, to anchor your introverted fantasy world. You don't need data or climate science to persuade your tricky mind. You just need to tune in and act.'

Tapping into imaginative thinking is not routine for all the knowledge cultures – yet advances in each culture are by creative leaps. Holistic thinkers are naturally creative, since they look for the essence or a synthesis of a group of ideas.

Revolutionary science is creative. However, normal science follows familiar paths. Organizations often seek to be creative, although they only too often constrain ideas to a pre-set mould. A creative community has an energy and sense of excitement that we are, of course, hoping to develop in this collective learning process.

Learning styles and creative thinking in Stage 3, Step 4

The mental shift from linear, logical, so-called left brain thinking to a more global intuitive mode is within the capacity of us all. However, those of us used to working in the logical, grounded mode may need a helping hand to make the shift. The case studies in Part 2 include the exercises used for change processes, ranging from an individual to an entire region.

A creative leap that generates new ideas can be encouraged through drawings, stories, songs, visioning and imaginative frameworks. Drawings can be individual and then brought together, or constructed as a group. Have your group draw the answer to the focus question. Imaginative scenarios are told as stories, not projections or predictions. Scenarios developed by a diverse group can be highly original and rich in ideas. For some groups who are already creative in their thinking, writing a haiku or a narrative poem together might release new ideas.

Visioning (guided mental imagery) can be very powerful with an experienced guide. Ideal outcomes to the focus question can be generated as a group (see Visioning in Part 3). Finally, a framework can be useful for focusing the imagination. Group members brainstorming the shape of a new organization or community can imagine it as an automobile or a bicycle (see Figure 7.1 for an example).

Get moving before the trees leave.

Source: Art of Moving (2006)

Figure 7.1 Staying ahead of the game

Step 4: Collective Ideas: What Could Be? 59

Making the connections between what could be and what can be

Participants in the learning cycle have made the leaps from right to left brain in going from ideals to facts in Stages 1 to 2. They have gone from left back to right brain in moving from facts to ideas in Stage 2 to 3. It is now time to move from Stage 3 to Stage 4. Stage 4 puts into place a practical program that answers the focus question by including the new ideas. David Kolb, who identified this adult learning cycle, was convinced that no learning was achieved, and so no change process possible, without completing the final stage of each cycle. That is, combining the new and the old ideas in a changed practice.

How often have you heard an inspiring lecture or been to a challenging workshop one week, and the next week can't remember what it was all about? The learning lies in the step of placing the new ideas into effect, no matter how outrageous they may seem.

A bridge between the new and the old ways of doing things can be created through identifying how you would know if the fresh ideas have been put into practice. This step sets the stage for choosing indicators for the monitoring and evaluation of the action plan that follows.

Tools:

See Monitoring and Evaluation in Chapter 17.

SALLY AND RICHARD 5

A roller-coaster ride

Reflecting on the first part of the collective learning cycle that guided the project, Sally and Richard again felt both excitement and anxiety.

Bringing the diversity of participants together in the room and having them develop a shared set of ideals (the potential goals for the project) created a real enthusiasm among participants.

However, the atmosphere in the room changed quickly when participants moved on to sharing their perceptions of 'the facts'. Technical experts and policy leaders in the room were not accustomed to the views of community participants and vocal individuals being given equal weight with their views. Good facilitation was needed to ensure everyone listened to and respected each other.

All was going fine so far, but Richard and Sally only had the lunch break to think of how they would establish the break between checking the facts and using the imagination to develop future options.

After lunch Richard led a visioning exercise in which participants were asked to draw what their new region would look like (see Visioning in Part 3) while Sally provided the drawing materials. After that, each table was asked to join up their pictures on one large sheet and name the connections between them. Explaining their pictures to each other generated another buzz of ideas. Naming the connections opened up a pathway to preparing an action plan.

This shared effort rekindled enthusiasm in the room, but now it was time to move to some real world action plans.

As she gained confidence in the process opening up before her, Sally commented on the roller-coaster ride from the creative leap to being strictly logical and back again.

8 Step 5
Collective action: What can be?

Summary: This chapter lists the practical actions to be taken to put new ideas into practice in the old environment

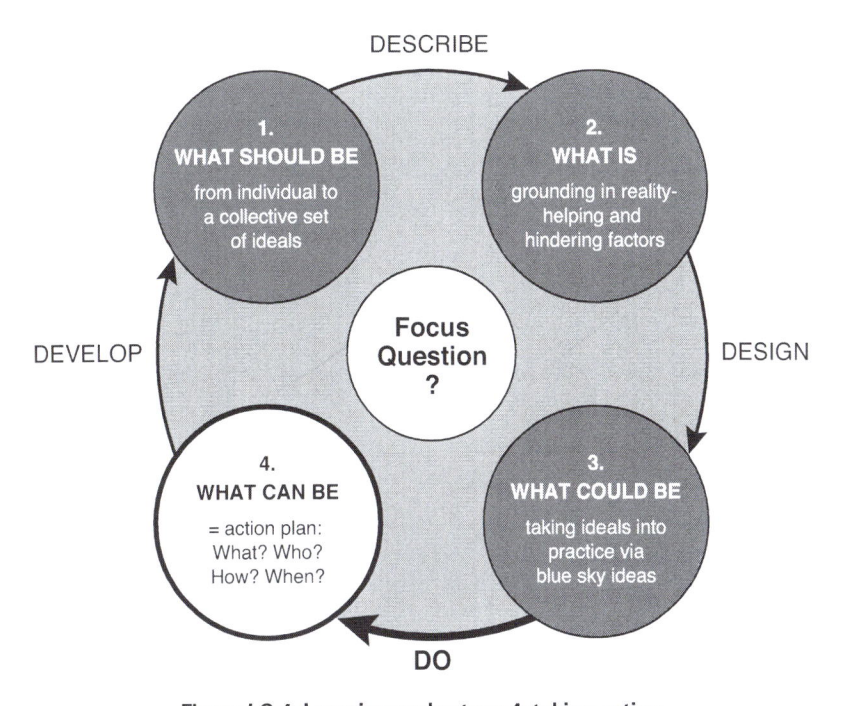

DESCRIBE

1. WHAT SHOULD BE
from individual to a collective set of ideals

2. WHAT IS
grounding in reality- helping and hindering factors

Focus Question ?

DEVELOP   DESIGN

4. WHAT CAN BE
= action plan:
What? Who?
How? When?

3. WHAT COULD BE
taking ideals into practice via blue sky ideas

DO

Figure LC-4 Learning cycle stage 4: taking action

Without losing sight of any of these crazy blue sky dreams, it's now up to the collective to put those dreams into reality – 'what can be'. This is not about going back to all those constraints. It's about making the dreams practical actions: determining who wants to be involved in what action and how, with a timeline of when it will happen, and indicators of what happens.

Step 5: Collective Action: What Can Be? 63

Each action plan needs a defined what, why, who, how, where and when, that is:

- What is the project/ program?   Project title
- Why is the project needed?   Aims
- Who is involved in the project?   Names of the project team
- How is the project to be resourced? Resources and in-kind support
- Where is the action to take place?   The setting
- When is the action to take place?   Timetable
- How do we evaluate the action?   Indicators

Examples of Action Plans can be found in each of the case studies in Part 2.

The collective learning team accepts responsibility for arranging (not necessarily doing) for the shared action plan to go ahead. One member of the team acts as catalyst for each project, until at least the next meeting – and possibly for months if that is feasible.

Once the plan is put into effect it is best if a plan member takes charge of regular action team meetings and reporting on the indicators. Their other role is to ensure that each follow-on meeting is positive, friendly and fun.

Making the connection between what can be and what actually happens

The close of this cycle is the beginning of the next. The action teams launched in Stage 4 start off their new program with sharing their ideals for their project, and so the cycle continues (Figure 3.2). The new project or program may be on quite another dimension to the one in place before the collective learning process. In one program a collective learning cycle began with the goal of improving collaboration and ended up as a shift in the sponsoring group to a change management program. In another case the initial goal was survival of a rural community short of water, but shifted to a community development program for the social cohesion of the region.

Kolb predicts that in any change process there will be no learning until the new ideas have been put into effect. In a collective learning process Stage 4 is often the first time a transformation becomes visible to the wider community.

Then the inevitable strains between new and old appear. While the action plan lays the ground work for the potential transformation, it does not end the commitment of the original team members.

This is the point where opponents come out of the woodwork. In one sense the validity of the transformation becomes a test of the degree of rejection of those not involved in the collective learning process. The circle will need to widen if the change is to be embedded in the surrounding society. This is just one more reason why the many interests are involved in the learning cycle from the beginning and are still involved, so that the circle turns into the spiral.

Part 2 offers case studies of the use of the collective social learning process. You can choose from the case study closest to your own concerns or use it to widen your field of operation.

The case studies explore transformational change through individual reflections, conference contributions, organizational policy, professional practice, strategic thinking, community problem-solving and whole-of community change.

SALLY AND RICHARD 6

We got there!

Sally and Richard were pleased with the outcome. The community had developed an action plan that was very different from where they'd started. They were on a pathway to transformational change.

Reflecting on what they had achieved, Richard stressed the need for both the community and the project team to continue the learning process, while Sally recognized that the ways in which they had learnt and the tools they used could be used in other programs.

The results had been sufficiently rewarding that Sally and Richard were keen to set up their own collective learning process with a team that can assist others.

9 Step 6
Following on

Summary: This chapter describes ways to ensure that the changes to implement the action plan are put in place.

Source: Art of Moving (2006)

When the workshop ends the need for a follow-through begins. This includes the implementing, monitoring and evaluation of the projects designed at the workshop as they re-enter the pre-existing environment. The enthusiasm generated in the workshop needs nurturing in a possibly hostile setting.

As we have already noted, there is no lasting change until the new ideas have been put into effect back in the pre-existing environment. With transformational change, the social environment itself will not necessarily welcome the change. This is where we move out into the next turn of the collective learning spiral.

There is a proverb 'you can't put new wine into old bottles' and that applies here. Luckily a social environment is not as fixed as a bottle. The new wine is the collective learning we have evolved in the learning cycle. The old bottle is how the society is used to functioning. A better comparison is to think of the collective learning moving back into the surrounding society more like yeast in bread-making than wine in a bottle.

What is needed now is to follow the project design established in Stage 4 of the cycle. This will not be as easy as it may have seemed in the enthusiasm of the workshop. That enthusiasm needs to be maintained through already full workloads and while moving into the previous environment with its own traditions.

The proposers of the projects in the learning cycle do not have to do all the work themselves. One mode of re-entry into the surrounding social environment may be for one member of the first learning cycle to run the workshop process again, this time with the new project team. A member of the first cycle could give a training workshop to the team leaders of a new project. This could continue into the next turn of the spiral, within the larger environment (see Chapters 12 and 14).

There are many and varied environments in which the action stage of the learning cycle has been effective. Case Studies 1 and 5 are examples of where the original collective learning spread throughout a capital city and a region, carried forward by further collective learning cycles in a ripple effect. Chapter 15 has examples of transformational change which reverse the failing environments of a lead-affected city and a dying town.

Recruiting the new members of the project teams from each of the key interest groups we have identified here (key individuals, affected communities, relevant specialists, influential organizations and holistic thinkers) will need further team-building. Examples of using the collective learning process for team-building, are found in Case Studies 13 and 14.

Tools: A multitude of tools can help start off a mutual understanding among the members of a new team, for instance 'speed dating' and 'world cafe'. Speed dating involves rotating the team members in pairs, with one-to-one sharing of skills and expectations and arranging future meeting times in a convivial space. This may be a cup of coffee once a week or an informal team meeting once a month.

In speed dating, half the participants stay still and the other half rotate around them, with 10 minutes for each conversation. A final plenary amalgamates the whole team's skills and expectations and puts together a meeting plan. The results of a speed dating exercise are reported in Case Study 6.

World cafe involves generating a set of questions about the project from among the whole group. These questions might include the goals, the resources and the timetable. Tables for four are set up, with one member staying at each table and the others rotating every 15 minutes. The questions are addressed at each table and the answers of the three rounds collated by the 'fixed' member. The collected answers are put to a plenary meeting, which then decides the future program as a group.

Both of these methods help fast-forward the program for the group to which all members have contributed. They are suitable for putting into practice a project that already has an action plan.

At the early planning stage the monitoring and evaluation of the program is set in place. In Part 2, Case Study 11 focuses on the use of the collective learning cycle to arrive at an appreciative, interactive and collective design for monitoring and evaluation.

In Chapter 5, What should be? the importance of confirming the collective learning outcomes by a timely report was emphasized. Planning for the report begins with this first question, not at the end of the workshop. The accounts of the 16 case studies in Part 2 were largely based on such reports, after the participants had confirmed the findings as valid for them. The circulation of the report to all parties that can help put the action into effect is still part of the learning cycle.

10 Guiding transformational change

Summary: This chapter recommends treating transformational change as a wicked problem with many dimensions. It goes on to review the multiple learning styles that are likely to be found when guiding collective learning. It then discusses the relationship between learning, knowledge and power.

Transformational change as a wicked problem

According to learning consultants, experiential learning is about creating an experience where learning can happen. How do you create a well-crafted learning experience? The key lies in the people who are the collective learning guide or guides, and how they facilitate the learning process. And while it is the learner's experience that is most important to the learning process, it is important not to forget the wealth of experience a good facilitator also brings to the situation.

An effective facilitator is one who is passionate about his or her work and is able to immerse participants totally in the learning situation, allowing them to gain new knowledge from their peers and the environment created. These facilitators stimulate the imagination, keeping participants hooked on the experience. One outcome measure is for the participants to ask 'When do we meet again?'

A fun learning environment, with plenty of laughter and respect for the learner's abilities, fosters an effective experiential learning environment. It is vital that the individual is encouraged to directly involve themselves in the experience, in order that they gain a better understanding of the new knowledge and retain the information for a longer time.

Chapters 2–9 developed a framework for bringing the knowledge cultures together in an open-ended learning spiral. The framework involved sharing the values of the different knowledge cultures; pooling their evidence; generating creative solutions; and acting on the ideas for that particular context. The next question must be: what are the responsibilities of the facilitator and leader of the change management process in ensuring the collective learning process successfully brings a hoped-for transformational change?

First, the guide needs to determine whether the task is a simple or a complex problem, and to be clear about the difference (Figure 10.1). When the goal is transformational change, not only will it certainly be a complex problem as in Figure 10.1, it will almost certainly be a wicked problem. A wicked problem is defined as one that can only be resolved by changes in the society that generated it. It is wicked in that it resists ready-made solutions and requires some hard thinking. The originators of the work on wicked problems, Rittel and Webber (1973) comment that, although problems such as these are not morally wicked, it would be morally wicked to treat them as simple problems with pre-determined answers.

DIFFERENCES BETWEEN SOLVING SIMPLE VS. COMPLEX ISSUES — PROCESS DIFFERENTIATION.

A. Solving a simple problem: using logic

1. Scope the problem (current)
2. Choose the solution (solution)
3. Test the solution
4. Learn from the test
5. Adapt solution to fit current problem
6. Finalize solution

B. Resolving a complex problem: using the imagination

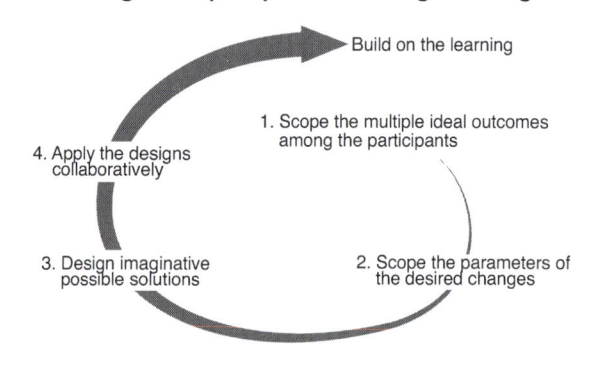

- Build on the learning
- 1. Scope the multiple ideal outcomes among the participants
- 2. Scope the parameters of the desired changes
- 3. Design imaginative possible solutions
- 4. Apply the designs collaboratively

Figure 10.1 Differences between resolving simple and complex problems

The characteristics of wicked problems are that:

1. Wicked problems evade clear definition. They have multiple interpretations from multiple interests, with no one version verifiable as right or wrong.
2. Wicked problems are multi-causal with many interdependencies, thereby involving trade-offs between conflicting goals.
3. Attempts to address wicked problems often lead to unforeseen consequences elsewhere, creating a continuing spiral of change.
4. Wicked problems are often not stable. Problem-solvers are forced to focus on a moving target.
5. Wicked problems can have no single solution. Since there is no definitive stable problem there can be no definitive resolution.
6. Wicked problems are socially complex. Their social complexity baffles many management approaches.
7. Wicked problems rarely sit conveniently within any one person, discipline or organization, making it difficult to position responsibility.
8. Resolution of wicked problems necessarily involves changes in personal and social behaviour, changes that may be strongly resisted or encouraged, according to circumstances.
9. Each wicked problem is unique in that it is a product of time, place and people; learning can be transferred from one wicked problem to another, but solutions cannot.

Since wicked problems are part of the society that generates them, they require the leader or guide to challenge the current understanding of the participants in the change. Rather than following the fixed trajectories of pre-existing solutions, addressing wicked problems involves the participants in exploring the full range of potential solutions. This asks for the guide to be tolerant of ambiguity and to be comfortable with paradox.

In particular it asks the guide to be creative, not only in planning the six steps of the collective learning circle, but in the process of putting it into practice. In the interactions among the diverse interests, many surprises emerge and the guide must be ready to steer those in constructive directions.

Rittel and Webber's work gives a practical foundation to the quip attributed to Einstein: 'You cannot solve a complex problem through the same thinking that created it'. Work on wicked problems has taken many paths, from the design of cutting-edge information systems, to engineering designs and systems thinking.

Each wicked problem rests on underlying paradoxes, that is, self-contradictory statements in which both propositions are true. To address the issues a society has not been able to resolve, that same society has to resolve to collaborate in social change. Paradox is embedded in transformational change. The co-existence of judgements of right and wrong; of fixed goals and trade-offs; of continuity and change; of the urgency for problem resolution and the acceptance of no final solution; and of competing organization and disorganization in the one society.

Wicked problems typically contain multiple ethical positions, multiple world views, and multiple ways of constructing knowledge, the three foundations of an open critical inquiry. In the inquiry into simple problems, the problem would be approached from one world view and one type of construction of knowledge, and expecting the two to be logically consistent. The ethical perspective would not usually be examined.

The shift from managing the traditional bounded inquiry to guiding collective learning is marked by a different construction of the task. No longer is the inquiry the sole responsibility of one specialist discipline or profession; it seeks the evidence from all affected parties. Nor are the findings of the inquiry expected to be final, certain or complete. A paradox is welcomed as offering a potential solution, not treated as an error to be eliminated. Ideals and ideas do not remain isolated in the head, but are linked to actions on the ground.

Moreover, collective learning asks a different type of question from one within any one of the specialist disciplines. The philosopher John Passmore differentiates between questions of ecology and ecological questions. In the first, an inquirer perceives their task as adding to the knowledge base of the discipline of ecology. In the second, the aim of the inquiry is to unravel a complex problem such as climate change that, while the trigger may be ecological, affects all aspects of the socio-environmental system.

The entry point for an inquiry into a wicked problem is usually the challenge of some wake-up call, a crisis event, a new idea, or a shift in social expectations. Since all parts of a complex problem are inter-connected, any entry point can open up the complexity of the whole. The guide's role is to identify the range of world views involved in a proposed transformational change in response to the challenge.

The world views of the interest groups might be of the planet as an inexhaustible source of resources or a plundered planet. It might be regarded as divided between Western wealth and Southern poverty, or as incompatible sets of technical and social issues. It may have been assumed that the state of the world will always be in a state of flux, or that a stable state is the desirable goal. It is up to the facilitator/guide to help the participants clarify their positions and make them transparent to each other.

The seminal work of Thomas Kuhn establishes that any knowledge tradition is determined by a combination of approved methods of inquiry, tests for truth and theoretical frameworks. Etienne Wenger identifies the way in which the practitioners working in any knowledge tradition form a community of practice. A guided transformational change includes and goes beyond community memory, academic disciplines, and organizational agendas. In accepting the equivalence of the different ways of knowing described in Chapters 1 and 3, a collective learning guide services a distinctive community of practice, one well-equipped to guide transformational change.

Identifying multiple learning styles

Over their lifetime everyone develops their own learning style. That learning style is the outcome of interactions among lived experience, social learning and personality. A second responsibility of a guide to transformational change is to match the tools they use and the activities they design to the learning styles of the participants. The Kolb process identifies four principal learning styles, which in turn lead to many other influences on the diverse participants: diverging, assimilating, converging and accommodating. Any of these influences is the concern of the learning guide.

It can be dangerous to apply past experience directly to a new group. Every group contains different people, at a different time, and a different place. Thinking of past groups can become a self-fulfilling prophecy, not a fresh change to make a difference. So think of the river flowing on – if you step in it today it is not the same river you stepped into yesterday, and it won't be the same river you step in tomorrow. This is not only true of groups. It is true of sustainability issues in a rapidly changing world.

We have already discussed in Chapter 1 how David Kolb found that particular occupations preferred one form of learning styles to others (see Figure 10.2). Researchers, scientists and engineers have a talent for convergent thinking, moving from the general to the particular. Artists, writers and poets display divergent thinking. Service professionals, such as in health and administration, are strong at assimilating, linking new ideas to an existing framework. Organizational managers and theatre directors are likely to be accommodating, changing the framework to adapt to the new ideas. Of course, everyone uses all of these skills some of the time. Some of the most spectacular advances in introducing social change have been the work of people who combined them all.

Accommodating	Divergent
Program managers Architects Theatre producers	Artists Strategic planners Writers
Assimilating	**Convergent**
Engineers Dentists	Physical scientists Accountants Technicians

Figure 10.2 Kolb learning styles linked to a range of occupations

Modes of learning for transformational change need to be inclusive, and to celebrate the diversity of the contributors. It follows that there needs to be an acceptance of the web of the different learning styles through which the learning is generated. Each interest group will have different roles, factions and interests within that group. For instance, a community-based conservation group member might come from either end of the public use continuum – total exclusion or public facility – or anywhere in between. The same goes for land development companies, government agencies and research groups.

A strong influence on collective learning is the various participants' approach to change in the first place. Although attitudes to change vary widely according to the particular type of change, most people have a consistent approach to change in general (Figure 10.3).

SPECTRUM OF APPROACHES TO CHANGE

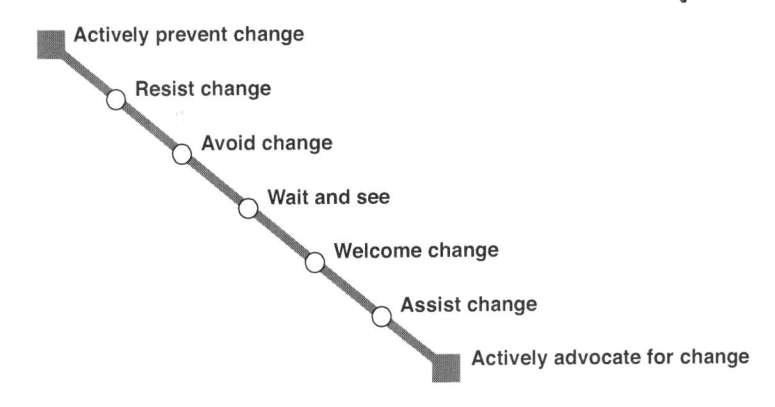

- Actively prevent change
- Resist change
- Avoid change
- Wait and see
- Welcome change
- Assist change
- Actively advocate for change

Figure 10.3 Attitudes to change

We know there are early and late adopters of new skills and technologies, and this also applies to transformational change. Whatever their background, people vary on a continuum from actively preventing change to actively advocating for change.

Guiding change through all the stages of the collective learning spiral can appear too complex for any one individual. Fortunately, our brains have a most powerful unifying capacity. Nobel Laureate in Neuropsychology Alexander Luria records how stroke- and accident-damaged brains – even if 50% of the brain is destroyed – can restore the identity of the person, and reunite with previous learning.

Advances in ideas have usually been the outcome of a collective mind, yet the diverse contributors to the ideas have not always been recognized. Denis Brian, in *The Voice of Genius* (1995), records interviews with 20 then-living science-based Nobel Prize winners. Individual characters though they were, they all, even Albert Einstein, spoke of the essential contribution to their discoveries of open-ended discussion with a few compatible spirits, not necessarily recognized as geniuses (Box 10.1).

> **Box 10.1 Einstein as a Collective Thinker**
>
> Albert Einstein was adapted for collective thinking biologically as well as intellectually. Having left his brain to science, scientists at first decided that it had the same structure and weight as other brains. Eventually, Sandra Witelson, an anatomist from McMaster University, found that the corpus callosum, usually a distinct rope of nerve fibres connecting the right and left hemispheres (the sources of our powers of analysis and synthesis), was a fused mass linking both hemispheres.
>
> Einstein reported that his capacity to publish ground-breaking articles which changed the face of physics from a mechanistic to a relative understanding of the universe in his early twenties was due to his discussions with a group of fellow students.
>
> Einstein's genius in conducting what he called thought experiments combined a vast grasp of mathematical structures with the capacity to make creative leaps. This arose at least in part from the two sources, close connection between the two brain hemispheres and the opportunities for collective discussion. Einstein had a greater physical and intellectual capacity for thinking synergistically, but the basic physiology and conditions of creativity are available to us all.

It would seem that collective thinking is as innate as thinking analytically. In guiding change, we need to remember that the present generation is recovering from ten generations of emphasis on logical analysis. As Michael Smithson (1989) comments, protection of the dominant mode of analytic thought has led to a tendency for social denial of the capacity of ordinary human beings to think synergistically, and to generate new thinking between them. This denial needs to be taken into account in guiding collective learning.

The Greek poet Archilocus in the 7th century BC used the analogies of the fox and the hedgehog to describe two very different ways of construing the world: 'the fox knows many strategies, but the hedgehog knows one great and effective strategy'. A study of the learning strategies of 750 undergraduates in the faculties of Arts and Science and a cross-faculty program revealed not two distinct learning strategies but three. One group, labelled holists (hedgehogs) were concerned with understanding the core of an issue. They were intolerant of complexity, but tolerant of ambiguity and uncertainty.

Another group, specialists (hounds) were interested in covering each topic in detail and in following each trail of inquiry to the end. They were concerned whether their contributions to knowledge were accurate and reliable. While they were tolerant of complexity, they avoided uncertainty and ambiguity. Generalists (foxes) shared with specialists the intolerance of ambiguity, and with the holists an interest in their learning being relevant. Hedgehogs, hounds and foxes can be linked to the learning styles of holists, specialists and organizations in the collective learning process.

The field of Gestalt psychology is concerned with understanding the capacity of the human mind to make sense of any complex information. Figure 10.4 presents the classic test, not of intelligence but of the personal mode of **synthesizing evidence**. Did you see the face in the figure immediately? After a few moments? Can you not see it at all? Studies confirm that about 5% of observers see the 'face' immediately, and after that cannot lose the image. About 5% will never see the figure, no matter how carefully it is explained to them. They will find this extremely frustrating. The other 90% will see it gradually, or after it has been pointed out. Everyone is intrigued by the task. There is no right answer. The figure is an ink blot.

Figure 10.4 Classic gestalt figure: 'Christ in the snow'

The 16-type matrix known as the Myers-Briggs Type Indicator offers an analysis of **personality types** that is frequently used in organizational development. The indicator cross-references the various ways in which people make their decisions, according to four dimensions: extraversion/introversion, sensing/intuiting, thinking/feeling and judging/perceiving. The personality matrix that results is intended to allow management to select a well-balanced team, well suited to collaboration. Each individual has the choice of extending his or her repertoire across the full spectrum, or making up part of a mixed team.

Howard Gardner, in his *Frames of Mind* (1983), challenged the orthodox assumption that intelligence is rational and objective. He identified **eight types of intelligence** and suggested there well may be more. The intelligences – Logical, Linguistic, Spatial, Interpersonal, Intrapersonal, Naturalist, Kinesthetic and Musical – have now all been the subject of field studies. Confirmed by many educational writers since, the types of intelligence have been unjustifiably allocated to certain occupations and personality types. In practice, we all use all of them all the time, even if to varying degrees.

Martin Buber, a German-Jewish philosopher of the 1930s, has reflected deeply on the relationships between the individual and the collective. Buber takes the position that each individual only exists through his or her interactions with other humans. These relationships he differentiates into **I-It, I-You and I-Thou**. The I-It relationship is an inheritance of the industrial era, when people began to use machines to undertake previously human tasks. The 'it' of machinery was transferred to the people who did the work – 'I-You'. The other relationship Buber calls the 'I-Thou', a bond between people which enlarges each person, and to which each person responds by trying to enhance the other person.

In research, there is a new drive to achieve a collective understanding drawing on the full set of nested knowledges described in Chapter 1. This multiple evidence base applies to both the thinking of the individual researcher, and to the subject of their research. The outcome has been found in practice to be more insightful, and closer to practical application, than any one of the knowledge cultures alone (Figure 10.5).

1. Thesis:
WHAT SHOULD BE?

2. Anti-Thesis:
WHAT IS?
e.g. Our cities are seedbeds of violence; the paradox is the need for both challenge *and* security

3. Thesis proposition:
WHAT COULD BE?
Theory: ideas on creating an ideal

4. Thesis inquiry:
WHAT CAN BE?
Practice: test ideas on the ground in collaborative action research

5. Thesis synthesis:
WHAT SHOULD BE?
Bringing together the threads of the study e.g. implications for design of a liveable city

Scoping five knowledge cultures' contributions on what should be

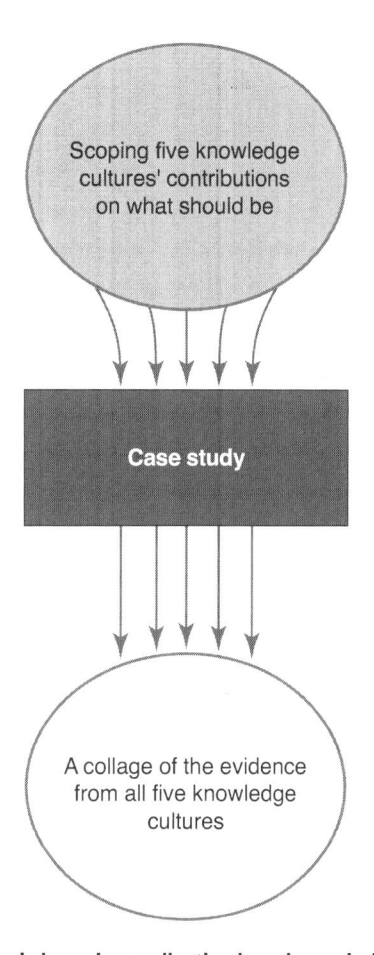

Case study

A collage of the evidence from all five knowledge cultures

Figure 10.5 **Model of a PhD thesis based on collective learning spiral**

Knowledge, learning and power

Chapter 4 explored the question of who to invite to a collective learning process directed towards transformational change. The conclusion was that, ideally, there should be equal numbers of participants from the five interest groups who could bring about whole-of-community change. The guide will need to make the basis of this choice transparent to everyone involved. This can be done through the diagrams in Chapter 4, informal discussion, or asking the group to complete a self-report page (Figure 10.6).

The Multiple Hats of Collective Learning

Five Roles	Structure	Basis for Learning	Significance
Private Individual		Personal Experience	For you?
Community Member		Stories Shared Events	For your community?
Specialist		Case Studies Observations	For your profession?
Organization		Agendas Alliances	For your organization?
Creative Thinker		Focus Creative Leap	For your imagination?

Figure 10.6 Questionnaire for participants in a collective learning spiral

All of us in modern Western society wear many hats – we fill many roles. As ourselves, we are individuals who develop our own ideas. As a member of a community, we have sympathy with the local experience. As a specialist, we have been trained in one particular set of skills. As a member of several organizations: voter, consumer, employee, manager... we can have divided loyalties. As a creative thinker, we make leaps of the imagination. Check out your own experience of wearing the different hats. Did you find conflict between the roles? See Figure 10.6 and ask the participants to complete the form for themselves. Case Study 10 contains the responses to these questions by the authors of this Guidebook.

In spite of recognizing that all individuals use all of the five knowledge cultures it takes to achieve transformational change, it is important to remember that most people only access one form of knowledge at a time (see Chapter 4). The guide can expect that key individuals will draw on their own experience, community members will draw on their local knowledge, specialists will draw on their own discipline, and participants from organizations will pursue that organization's agenda. Creative thinkers will come from left field. An important task therefore is to help the participants learn to draw on all their capacities for constructing new knowledge, which, of course, is one definition of learning.

A satisfactory definition of knowledge is notoriously difficult. Philosophers of the scientific era have labelled it 'justified true belief'. The 20th Century father of the philosophy of science, Karl Popper, proposed that knowledge emerges from a combination of the field of ideas, the bio-physical world and the eye of the observer. It is clear that knowledge is a complex, dynamic and open-ended construct distinct from information, in spite of the continuing confusion between the use of the terms knowledge and information.

A complicating factor in increasing each person's informed access to all the knowledge cultures is the power imbalance between them. Specialized knowledge has been long regarded as superior to all others, more accurate and more reliable. More recently specialized knowledge has been challenged by organizational knowledge, where political or industry strategic thinking are placed ahead of expert advice. Climate change is an example. Community knowledge is less likely to be accepted, and individual knowledge dismissed as biased. Creative knowledge can be dismissed entirely until it becomes dramatically significant, as in the examples in Chapter 4.

Equalizing the power imbalance requires an understanding of the way in which knowledge is constructed, since knowledge is only too often confused with information. It is helpful to consider the chain made up of data (items of information), information (the arrangement of data), knowledge (the arrangement of the information) and wisdom (a very rare capacity for applying knowledge). This chain of events is discontinuous; one does not necessarily lead to the other.

Given the many dimensions of thought and the wide range of tools that contribute to the construction and sharing of knowledge, it is not surprising that knowledge is shaped differently in different societies, during different eras, and by different communities. Knowledge can be considered a commodity that can be bought and sold, and formal and informal education can be approached as a knowledge industry. Our capacity to reflect, connect, inquire, and remember suggests that knowledge sharing is far more complex than that.

We can imagine the various approaches to knowledge and learning that a learning guide could take. At one end we have those who consider that knowledge is generated in the mind of each individual. The individual stores the knowledge in a mental schema, and their learning is slotted into that schema. This view is held within cognitive psychology. At the other end of the learning continuum we have the social learning theorists who understand each person's knowledge as shaped by the cultural givens of their society. Learning is then understood to be a social process, which can only occur when there is a significant change in enough members of the society.

Once again, most practitioners in the field are well aware of the reciprocal nature of the individual–social learning continuum and position themselves somewhere along that continuum, according to the context. The perspective that best supports collective learning lies in the fourth quadrant of Figure 10.7.

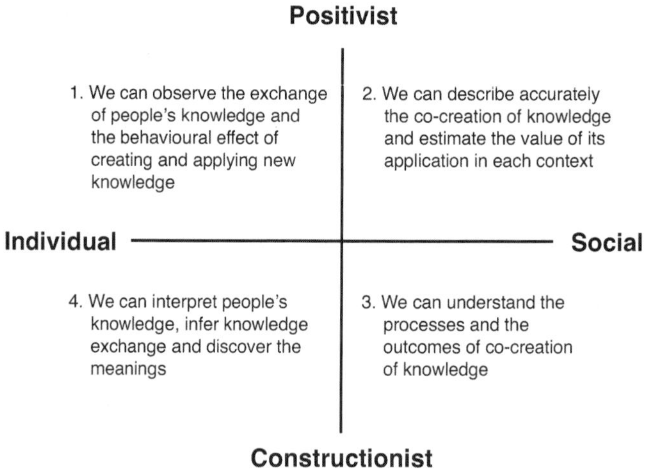

Positivist

1. We can observe the exchange of people's knowledge and the behavioural effect of creating and applying new knowledge

2. We can describe accurately the co-creation of knowledge and estimate the value of its application in each context

Individual ——————————————— **Social**

4. We can interpret people's knowledge, infer knowledge exchange and discover the meanings

3. We can understand the processes and the outcomes of co-creation of knowledge

Constructionist

Figure 10.7 Interaction between perspectives on learning (le Borgne, Brown and Hearn 2011)

Conclusions

The collective learning process can seem complex and confusing, even to an experienced facilitator. Even more confusing is that the spiral does not follow some of the more usual approaches to guided learning. In collective learning for transformational change, values come before facts, there are multiple legitimate constructions of knowledge, the use of imagination is an essential step, and collaboration is achieved only after a mutual learning process.

This chapter follows Chapter 1 on the theory of the collective learning cycle, Chapter 4 on the multiple knowledges required to permit transformational change, and a practical account of the six steps of transformational change. The aim has been to take account of some of the complexities that are inevitable in the collective learning process, such as the type of problem, the range of learning styles, and the relationship between knowledge and power.

These topics have been treated here as simply and directly as possible for easy access to the ideas. However, where a collective learning facilitator/guide finds that any one of the topics in this section is particularly significant or challenging in the course of their work, it should be possible for them to expand on the topic through the references in the bibliography, the tools in Part 3, or an internet search engine.

We wish them the best of luck with an exciting and rewarding enterprise. Part 2 which follows has 16 examples of making transformational change a celebration of what could be.

PART 2.
CASE STUDIES

Celebrations of Collective Learning

11 Holding the party

Summary: This chapter explores the reasons for treating a collective learning process as a celebration. Sixteen case studies illustrate examples of deliberate transformational change.

Overview

Why celebrate? Because there need be no more dreary meetings. No more boring tasks. Think about combining a set of diverse interests as if you are holding a party. Tap into the excitement generated through pooling skills in a constructive collective that works together for a transformational change.

The party: Holding a good party is not a simple matter. Success depends on a shared goal (reason for coming to the party), a creative focus (party theme), sympathetic leadership (party hosts) and effective organization (so party events go smoothly). As with a party, you know your learning process has been a success when the guests enjoy each other's company and look forward to the next time.

The case studies: In Part 1 we commented on the wide range of possibilities for establishing collective learning. Part 2 offers eight pairs of case studies of transformational change enabled by a particular approach to celebrating collective learning (Figure 11.1). Each case study is matched to the form of celebration that fits it best (Table 11.1). After reading this chapter on the overall shape of the case studies, we suggest you move to the type of celebration closest to your own interest, and then cherry pick among the others.

Connecting threads

These 16 case studies were conducted at a time of general concern about achieving transformational change towards a just and sustainable future for humans and for the planet. Linking matters of health and of environment have come to be regarded as central to this goal. The transformational changes described here have been initiated by individuals, local communities, specialists in the relevant fields, government and non-government agencies and creative thinkers. Each initiative addresses a different future fashioned by these five sets of interests working as a collective (see Part 1, Chapter 3).

The task changes competition (if you get more, I get less) to collaboration (we can all share) and then to a collective transformation (we all get more than we ever expected).

Three threads run through every one of the examples of transformative collective learning which make up Chapters 12–19. They are: setting up a decision-making collective, the sequence of the experiential learning stages, and an open leadership style.

Figure 11.1 Finding the whole picture

Decision-making collective: the collective for each collective learning cycle is made up of the full set of interests who can bring about transformational change in any community. A full decision-making set consists of key individuals, community members, relevant experts, influential organizations and creative (holistic) thinking (Part 1, Figure 4.2).

The scale may be global, national, regional or local. The sense of community can come from sharing a place, a profession or an issue. Experience has shown that the full collective set is needed for any significant change. Therefore the full decision-making set needs to be invited to every party.

Learning stages: collective learning takes place through guiding a group through a cycle of six essential steps, each cycle being a turn of a collective learning spiral: 1. setting the scene, 2. developing individual ideals (What should be), 3. describing multiple realities (What is), 4. designing a different future (What could be), 5. doing – putting the design into practice (What can be), and 6. establishing transformational change (Part 1, Figures 3.1 and 3.2).

The goal is not consensus nor is it to rank ideals in priority order. It is to learn from difference. Each turn of the collective learning spiral starts with ideals for the future, and only then moves to consider the facts. The steps reverse the usual practice of facts before values, and the order often needs to be defended.

Leadership: transformational change through changing from individual to collective interests is not a magic bullet. It takes time, care, an appreciation of diversity and a welcome for new experiences. Changes can be driven by anyone in the proposed collective, an individual, a community, a specialist, an organization or through a creative leap. The goal for everyone, especially the guide, is to learn from each other.

The guide can be one person, a whole team, an inspirational idea and co-leadership. Leadership may be applied through facilitation, chairing, administration, or knowledge brokering, best though is a team of all four. The leaders ensure that all interests contribute to the learning process on equal terms, and that they enjoy the process as well as the outcome. The content of each of the boxes in the learning cycle is developed by the members of the collective during the learning cycle and so cannot be determined beforehand. This can be somewhat unsettling for those who are used to working with predetermined aims and set objectives.

Table 11.1 What sort of celebration fits which sort of change?

Type of party	Type of change	Case study 1	Case study 2	Case study duration
Ch 12 Bon voyage	Managing whole-of-community change: *an open-ended journey*	Community-based change	Organization-based change	Three years, continuing
Ch 13 Cocktails	Introducing new ideas: *a contribution to an existing program*	Fresh ideas for a conference	Fresh ideas for a strategic plan	Three hours
Ch 14 Opening night	Initiating long-term change: *breaking new ground*	Starting from scratch	Making a fresh start	Three years then withdraw
Ch 15 House-warming	Changing problem communities: *a fresh start*	Polluted town	Dying town	Six months
Ch 16 Coming of age	Achieving collective thinking from individual knowledge	Lateral and linear thinking	Indigenous thinking	Two hours
Ch 17 Street party	Monitoring and evaluation: *getting to know the neighbourhood*	Project evaluation	Sustainability reporting	For length of the project and beyond
Ch 18 Bring a plate	Teamwork: *diverse contributions to a successful outcome*	Textbook, co-authored	Textbook, edited papers	Two days plus a half day
Ch 19 Going it alone	Working from the guidebook in other cultures	Independent school, Bali	Government agency, Ethopia	Two months

About the case studies

The case studies offered here have been selected from over 300 examples of the use of the collective learning spiral as a vehicle for transformational change. In every case the change process is built on the six steps of a collective learning cycle in Part 1, Figures 3.1 and 3.2. The nature and direction of the change is always determined by the collective itself.

Ideally each collective working towards a transformation is made up of key individuals, members of the affected community, relevant specialists, people from influential organizations, and creative thinkers. This mix is not always attainable, so each cycle starts with as many as possible.

The emphasis in each case study is on moving towards a more just and sustainable future. However the examples of collective learning process that follow can be applied to any complex issue.

Each example is introduced here by the style of party that matches the type of transformational change. Examples are then summarized by:

- the story
- the type of transformational change
- the form of collective
- the use of the learning spiral
- the leadership style
- the context
- the outcomes of the four learning stages
- the follow-up

This part of the book offers eight different ways of celebrating collective transformational change, with examples from two different situations for each one. Readers can choose which example best matches their own situation (Table 11.1).

Each case study contains a particular exercise that has been found to assist transformative learning. Examples are: creative thinking in Case Study 1, critical friends (Case Study 4), keeping a learning journal (Case Study 5) and self-testing for one's own knowledge culture (Case Study 9).

The stories of transformational change include long-term change for a city as a long and exciting journey. When friends start off on a long adventure they hold a Bon Voyage party (Chapter 12).

A community in real trouble can just fold up and die, or its members can re-think who they are and what they could be. Communities can turn inwards to change themselves, or look outward to find new opportunities (Chapter 15). In either case, the effect is very like moving to a new house – finding a new beginning and a new way of life. A House-warming is then the appropriate style of party.

Collective authorship may take the form of a co-written book or a collection of linked papers (Chapter 18). In both cases the task is like a 'Bring a Plate' party where everyone brings a contribution to a meal. The result is a richer meal than anyone could cook by themselves.

Note: The postcards which illustrate this book are the product of a collective learning project the Art of Moving. They are all produced by students of the Design class at the University of Canberra.

SALLY AND RICHARD 7

Walking the talk

Background. Sally, Richard and their team leaders have run one cycle of the collective learning spiral for practice (Figure 3.1). Now they are getting ready to use their learning from that experience in their major regional change program (Table 11.1).

Setting the scene:

Although the initial collective learning project had gone well, Sally and Richard were still feeling a little daunted at the prospect of applying the process to a major regional change program.

Both felt they, and their team, needed some more experience in using the approach before taking it out to such a big program.

A session involving the whole regional change project team was set up to gain that experience.

As always, a focus question set the scene and provided a touchstone for the process.

Focus question:
How can we best act as guides in a collective learning strategy?

Results of a half-day workshop, held over lunch in a pleasant environment, were:

Ideals: What should be?

Answer (individually): Clarity of purpose and shared interest in the outcome; ensure participants are clear about what they are there for and have faith in the process; team members need mutual respect and honesty about personal aims for the workshop, and clear lines of responsibility.

Facts: What is?
Answer (individually):

+ve (supporting factors): A climate of creative imagination, and hopefulness; buzz of exciting new ideas; people profoundly catalysed to think.

–ve (hindering factors): The process used as a replacement for action; individuals angry at having to share their mental space; people not confident to express a range of creative, different, 'way out' ideas or making unusual connections.

Ideas: What could be?
Answer (collectively): Each knowledge culture needs to move from conflicts of interest to trust and cooperation. Essentials are the rules of dialogue, a party ambience, and mutual respect. Accept that participants are likely to be competitive, individualized and alienated. Provide a creative experience as a trigger for making the game change.

Action: What can be?
Answer (collectively): We can be a fantastic team, each working from our own skills base and at the same time in a collective team process. Together we can bring change. We need to find ways to share our collective learning circles, it won't just happen.

As Sally and Richard debriefed at the end of this workshop, each felt more ready to embark on a collective learning process for regional change and each felt more strongly part of a team committed to the process.

12 Managing whole-of-community change: Bon voyage

Summary: In this chapter, the collective learning spiral is used to guide long-term /transformational change in two different cities – one regional, the other a capital city.

Background: The two case studies in this chapter each follow three turns of the collective learning spiral as a springboard for ongoing change throughout a city. The transformational change is for a capital city in the first and a regional city in the second. In each example the spiral is treated as one long Bon Voyage party, sending the project teams happily on their own way. Each of these examples follows the collective learning spiral until the learning is embedded in the community and the community run the change process for themselves.

Case Study 1 follows three years of a change strategy for connecting environment and health in a capital city. The transformation strategy is initiated by a community action group. The transformation's long term survival depends on continuing community support.

Case Study 2 describes the first five years of a long-term strategy towards a sustainable future for a regional city. This time the project is initiated and maintained by a government agency which recruits the support of community and industry.

Case Study 1
Community-based long-term change

Project: Action for Sustainability and Health

Story: Small seed funding generates ripples throughout a small city

Transformation: From conflicts of interest among local decision-makers to community-wide collective actions

Collective: 20–30 local not-for-profit non-government organizations from industry, health, environment and the arts

Learning: Three turns of the collective learning cycle followed by ongoing programs throughout the city

Leadership: Four team members of a community organization: human ecologist, ecologist, teacher and change manager.

First turn of the collective learning spiral: the start of a long journey

Context:
A community group applied for a local Public Health grant of $30,000 as seed funding to develop an Action for Sustainability and Health program for their city. Program aims were to:

- develop a whole-of-community strategy for making the connections between sustainability and health in the city and its region
- establish a consortium of advocates for sustainability and health in the city
- initiate a snowball series of ongoing projects furthering sustainability and health for the city.

Setting the scene:
The four hosts of the party chose 30 likely leaders from the five decision-making interests in the community. Each had a track record of community action. A mix of key individuals, community activists, relevant experts, renewable industry interests and artists were invited to a collective learning workshop at 6–9 pm in the local Council chambers. The guests were potential 'door openers' to other existing groups and their possible projects. Refreshments were served, the Council setting was impressive, and the project team acted as hosts to welcome and see to the comfort of the guests.

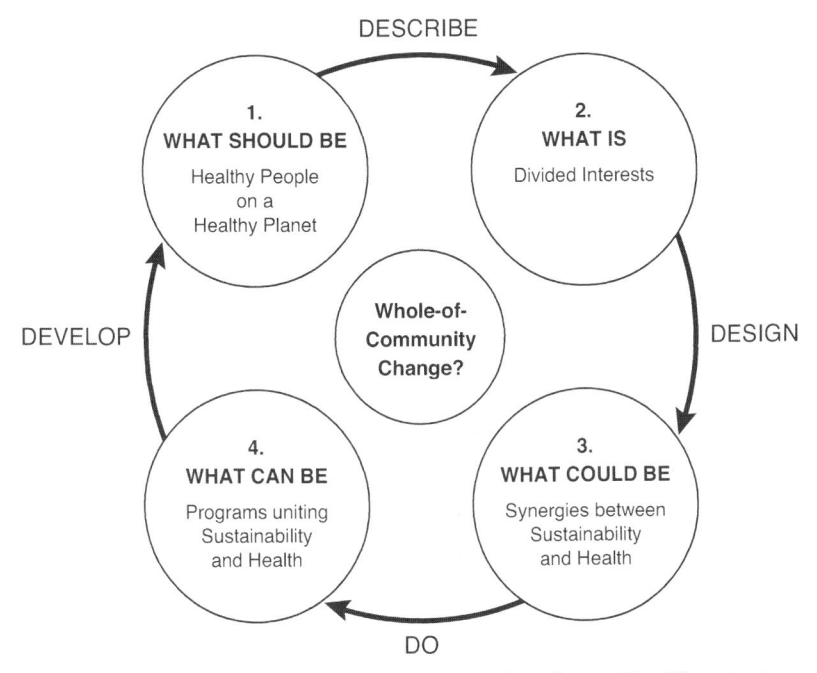

```
                    DESCRIBE
           ┌─────────────────────┐
           │                     │
      1.                              2.
   WHAT SHOULD BE              WHAT IS
   Healthy People           Divided Interests
        on a
   Healthy Planet
                    Whole-of-
  DEVELOP           Community          DESIGN
                    Change?

      4.                              3.
   WHAT CAN BE                WHAT COULD BE
   Programs uniting          Synergies between
   Sustainability              Sustainability
   and Health                   and Health
           │                     │
           └─────────────────────┘
                       DO
```

Figure 12.1 **Collective learning cycle for sustainability and health project**

The **focus question** on which the whole group decided as their first step was

How best to progress whole-of-community change towards the combined goals of sustainability and health?

A working definition of sustainability from the 2000 World Summit on Sustainable Development is 'the outcome of reconciling economic development goals, social needs, and the use of ecological resources, at the local and global scales'.

Following the four stages described in Chapters 5–8, workshop members answered the four questions in turn: What should be? (Ideals); What is? (Facts); What could be? (Ideas); What can be? (Action).

1. Ideals:

Summary of ideals: Healthy people on a healthy planet. The Action for Sustainability and Health project should not stand alone, but should contribute to a worldwide movement towards collective thinking for a connected world. The city should become an active, collaborative, networking, more peaceable community, using its wisdom and knowledge capital to build a just, sustainable and healthy society and city.

2. Facts:

+ve (supporting factors): The chief employment in the city was public service and higher education. The city had the highest average income and level of education in the country, hence a strong potential for learning about and supporting change. Over twenty organizations were working separately in sustainability and in health. If recruited for this project the organizations could provide community-wide resources.

–ve (impeding factors): The city had the highest consumption rate for goods and services, including food and transport, in the country and therefore the highest emission of carbon dioxide per person. There were very few links between the organizations, and they competed for funds and members. Health risks reflected the diseases of affluence in a well-to-do city: heart disease, lung disease, obesity. The matching environmental factors were lack of exercise, pollution and fast foods.

3. Ideas for a different future:

Potential projects that emerged during a brainstorm were:

- Social responses to the health issues of recycled grey water
- Innovative health pathways for Indigenous people
- Better diets for young people
- Assisting students to undertake assignments with the project
- Capacity-building process for young people as social change agents
- Establishing a regional environmental learning precinct
- An environmental focus for business through collaborative arrangements
- Theatre in schools

4. Actions:

Members of the initiating workshop agreed to pursue the potential of the ideas for a different future. They committed themselves to recruiting projects from within their own networks.

5. Follow-up:

The participants agreed to help organize a second collective learning cycle to bring together their proposed project teams.

Second turn of the collective learning spiral: 10 projects

Setting the scene: Invitations were sent to 60 people nominated for potential projects by the gate-openers from the first learning cycle. Fifty people accepted, and agreed to work toward answering the focus question:

How can we best design a practical whole-of-community change project in our own field of interest?

The potential project teams worked together in groups of seven for a two-day workshop in a pleasant garden setting arranged by the hosts, and paid for from the last of the seeding grant. Participants shared their ideals for a particular project, and identified the positive and negative factors for reaching those ideals.

After lunch, the performance art of a living sculpture left the participants ready to brainstorm ideas on how their projects would run. At the end of the day, 10 project teams had formed for possible follow-up action. Each agreed to seek out their own people, resources and outcomes for their project, with the support of the original host team.

The design of each potential project completed the learning cycle of ideals, facts, ideas and action. The resultant projects were (with host organizations in brackets):

1. Sustainability Scinema: Youth Film Festival on happy healthy sustainable living for youth (Healthy Schools program and local art group members)

2. Community Gardens: food and plant production, with a focus on nutrition, and no waste, using grey water, at Curtin gardens (Bushcare, Rotary, Uniting Church, Catchment Management Committee, water recycling industry)

3. Westhaven Concerned Citizens: an environmental learning precinct for Westhaven: community demonstration and tour for a natural bushland experience (Westhaven community development committee, school parents and friends, Landcare groups)

4. Indigenous wellness pathways: an Indigenous-designed Wellness Centre (Steering committee of 10 local Indigenous leaders, pilot study in rural town)

5. City buses as a vehicle for creative drama on sustainability (publicity officer, local drama groups)

6. Indigenous foods adventure (YMCA, Canberra nutritionist, conservation groups)

7. Art of Moving: Art display demonstrating reduction of energy consumption and maximising exercise. (Art School, university, schools)

8. ActWISE: Recruitment and design of a youth-led program for advancing sustainability in the region (Environment and Sustainability Centre, and university students)

9. Sustainable economic development strategy 2005-10 Models of good behaviour e.g. grey water, MECU financial products (Rotary, Credit Union, review of sustainability economic strategies elsewhere)

10. Plan for Living – in a nurturing environment. An educational complex for a wetland in the heart of the city (Nature and Society Forum, environmental educator, illustrator)

Step 6. Following on

Each of the projects created in the second cycle of the learning spiral was championed by the organizations that nominated them. Projects 2, 6 and 7 went into action immediately. Action plans for Projects 1, 3, 4, 9 and 10 sought funds and in-kind support for another round of the collective learning spiral. These projects then developed a life of their own. Overall, $600,000 in further funding and a great deal of in-kind help was generated from the first learning cycle funded by the seeding grant of $30,000.

Third turn of the collective learning spiral: six projects

Six of the projects generated a collective learning circle of their own as the basis for recruiting a wide-spread level of support. Westhaven Concerned Citizens, ActWISE and the Art of Moving Projects are examples of the follow-on from the expanding circles generated by the second turn of the Sustainability and Health spiral.

Westhaven Concerned Citizens

Ideals: To engender a shared community interest in and commitment to sustainable development in Westhaven.

Facts:
+ve: Planning completed; contacts with local community groups being established; a survey of the community and a community forum planned; and a community walk across the area ready to go.
–ve: A small group of people were doing all the work.

Ideas: A strategy whereby sustainable interest in and commitment to this community were strong enough for them to become a natural part of the local planning and development processes.

Actions: Support for the projects was recruited from Catchment Management and Landcare groups; local businesses and shops; media such as local radio and the local newspaper; links to government planners, politicians, community groups, websites; and links to the other projects.

ActWISE: Youth Leadership for Sustainable Consumption

Ideals: To facilitate, build and resource a thriving network of young people who act to create a more healthy, fun and sustainable city.

Facts: A pilot of 2 workshops, 50 people, tested training materials. Evaluation of the pilot taught the need to consolidate learning and to secure 5 year funding to expand coalition and networks.

Ideas: Potentially, run 4–8 workshops a year, facilitated by young people who are creating sustainable actions in their own lives. Offer peer leadership and support within the network of community, business and government.

Actions: Forge project links to Scinema, Western Region, Art of Moving and SEE Change to support youth action.

Source: Art of Moving (2006)

Figure 12.2 Sample postcard from Art of Moving project

Ideal: Design students at the local university were given the assignment of designing postcards to market an ideal: people abandoning pollution-emitting cars and improving their health at the same time by physical exercise.

Facts: Carbon dioxide emissions from motor cars causing climate change; obesity epidemic from lack of exercise.

Ideas: See Figures 12.2 and 12.3 for examples of 50 postcards.

Action: Postcards were put on buses, in shops, on billboards, in schools and sent by mail.

Source: Art of Moving (2006)

Figure 12.3 Two of the fifty Art of Moving postcards on replacing cars with physical exercise

Fourth turn of the collective learning spiral

Three years on, the founding community organization ran a collective learning circle to review all the continuing projects. Six of the original twelve projects had become an established part of city life. A further four projects had contributed new ideas. All project teams agreed that stimulating change in a city and its region had taken a combination of art, science, and strategic thinking at every turn of the learning spiral. All the project teams agreed that there was a buzz of excitement in working with diverse interests to bring about innovative social change. The outcome of the initial collective learning cycle with members from the full set of community decision-makers was an agreement to sponsor a further learning cycle of potential projects. In a second turn of the spiral the potential projects firmed up 12 project teams. Each of those teams ran their own form of the collective learning cycle to produce the ideals, facts, ideas and actions for their transformational change projects. Of the projects in the third cycle, six drew in over $600,000 in funds and much more in-kind support. The resulting network spread across the city and formed a fertile base for continuing change.

The founding community group continued to work together to build on their own learning from this spiral to further change projects. The case studies that follow have built in some way on the learning from the first case study. The spiral continues.

Case Study 2
Organization-based long-term change

Project: Community-wide Sustainability Services

Story: A transformational change program converts a city to a sustainability agenda

Transformation: From conflicts of interests to collaboration among community decision-makers

Collective: All agencies, communities, experts, industries and citizens of the city

Learning: Three turns of the collective learning spiral

Leadership: Visionary leadership from within a government agency

Context: The city of Beaconsville is a thriving port. A population of 300,000 live in a fragile ecosystem with three major industries, mining exports, tourism and sugar. The tourism drawcard is one of the world's major tropical reefs. Toxic chemicals used on sugar cane farms pollute the estuaries and so the reef. Minerals exported through the port leave dust over the town. Almost all the residents of Beaconsville are employed in one of these three major industries.

An ambitious program for a Sustainable Beaconsville took a great leap forward in March, 2007. The manager of the town Council's Environmental Services had already initiated a sustainable city program, attracting the collective interests of business, government and citizens. Existing Council sustainability initiatives were: partnership in an industrial Solar City, schools' Green Tree Ants and an environmental education program which followed a transect from Creek to Coral. A Sustainability Summit was organized to build on the success of these projects.

The Sustainability Summit was initiated by the deputy mayor and the director of environmental services. Their goal was to recruit all parts of and all interests in Beaconsville and its region in working towards a community-based collective action plan for a Sustainable Beaconsville. The intention was to mainstream sustainability principles throughout Beaconsville's ways of living and doing business.

Box 12.1. Report of an International Review Committee

Becoming Part of the Living System of Beaconsville

The Smarter Cities Challenge team experienced a packed agenda of tours, workshops, and gatherings - essentially becoming a part of the living system called Beaconsville. The City uses an approach to encourage citizen participation and collaboration called Collective Social Learning. The IBM Team experienced it on a grand scale when 180 people from across Beaconsville city and regional government, business, academia, not-for-profit and local interest groups came together to explore what should be, what is, and what could be.

This seemingly simple approach is incredibly powerful in the way it allows for different perspectives, wants and needs to be expressed, probed, and cultivated by small groups of people working together. By the end of the workshop, each individual commits to do one thing to encourage change, in our case for sustainability.

Our involvement in this workshop was important for multiple reasons: 1) We heard from a broad spectrum of Beaconsville stakeholders. 2) We were part of the process and co-created new ideas with the Beaconsvillians. 3) We walked away with a better understanding of how people energy is created and individual and collective action is achieved within the community. 4) Each IBMer personally committed to doing specific things to further sustainability.

The Beaconsville Difference

We observed that the people of Beaconsville have a passion for community betterment, innovation, and self-reliance. Champions in all levels of government, education, utilities, and industry showed strong evidence of willingness to brainstorm, cooperate, commit and act to change.

The strategies defined in multiple planning documents, and in workshops conducted with inclusion of the review team during our project revealed vision beyond the expected scope of a town of 190,000 citizens. Beaconsville's city government and its extended community are committed to becoming leaders in tropical sustainability.

First turn of the collective learning spiral: a sustainability summit

Setting the scene: The proposed Sustainability Summit using the collective learning cycle was directly preceded by two events: a Beaconsville City Council community participation forum, Beaconsville Have Your Say, and a Beaconsville Enterprise Business Leaders Forum Breakfast sponsored by industry. The aim of these meetings was to focus the thinking of those who would subsequently attend the Sustainability Summit.

The Summit was planned as a three day workshop with over 150 Beaconsville citizens from community, businesses and government. Everyone who accepted the invitation to the Sustainability Summit was asked to help decide on a focus question.

What collective actions can the corporate sector, experts, Council and the community take to progress a Sustainable Beaconsville?

The forum process began with three keynote speakers from engineering, business and action research, who focused the participants' minds on the key question. The workshop then followed the stages of the collective learning cycle.

Participants were seated in twelve tables of 10 whose membership was a mix of individual change agents, community members, technical experts, industry and government and creative thinkers. Each table had the diagram from Part 1. Figure 3.1 in the centre, A3 size. The task was for table members to arrive at a collective set of answers (not consensus, and not the lowest common denominator) to each of the four questions.

Each group started with the question 'What should be?' (What principles guide progress towards a Sustainable Beaconsville?). There was approximately an hour to summarize their answers on a Summary Sheet, all of which were then posted on the wall.

Groups then reviewed each other's report sheets, often asking the authors to clarify what they meant. The facilitator then summarized the main ideas/themes from each of the 12 groups, presenting them back to the whole group, for comment and to ensure all major themes were noted. This then created one document that summarized all 12 groups' collective answers.

This process was then repeated for the following two questions 'What is?' (What is happening now that either supports or blocks progress towards a Sustainable Beaconsville?) and 'What Could Be' (What could be done in innovative projects in Beaconsville now?). For this third question a short list of the projects was created from the 46 potential ideas. Each table then took one key idea and constructed an action plan.

Ideals: Examples from some of the tables

- A comprehensive road map for Beaconsville 'joining up all the dots"
- A communication hub which provided up-to-date practical information for the whole community
- Collaboration between citizens and research institutions for two-way information flow
- Respond to local and global changes that have already changed the cost/benefit equation
- Change from it being too expensive to incorporate sustainability strategies, to being too expensive not to use them
- Take advantage of the short space in time left for effective changes to happen, changes that can be compared in importance with the steam train and the automobile.

Facts:

-ve forces:	+ve forces:
individuals living in an isolated region	individual collective learning
tourism, mining, and academic industries	unifying sense of place
incompatible specialized solutions	collective action research
competing organizations	collaborative projects
absence of holistic focus	focus: a sustainable Beaconsville
unfamiliarity of collective thinking	enjoying collective thinking

Ideas: The ten tables identified 26 potential projects, of which ten went forward to action. Three of these are described below.

Second turn of the collective learning spiral

From ideas to action. Project sponsors agreed to meet and run a collective learning cycle for their project. Three examples of completed projects follow.

a. Structural change to Council organization

Ideal: To create a local focus for whole-of-city policy and practice towards sustainability in practice.

Facts: Resources needed from Beaconsville City Council, external professional agencies, other governments and academic/research organizations, and community groups. Timeframe – up to a year.

Ideas: A policy position in the Mayor's office of sustainability and support for a Tropical Centre of Excellence to provide advice and support for practical sustainability.

Action: Prepare briefing papers for Mayor and next Council meeting in 3 weeks.

Following on: A Sustainability Office was created within the Mayoral unit of Beaconsville City Council to develop and supervise new policy directions. Unusually, the office maintained the responsibility and resources allocated to the Environmental Services of the Council, now renamed Sustainability Services. Plans were put in train for the Office to develop a functional demonstration site including commercial/residential zones.

b. Expert technical change to solar thermal power

Ideal: A whole-of-Beaconsville sustainable energy balance.

Facts: Technical expertise and sponsorship from energy providers, solar thermal providers, developers, Council, Beaconsville Enterprise. Timeframe – Immediate.

Ideas: Solar thermal technology in the early commercialization stage, to ensure Beaconsville has a renewable energy source on an industrial scale.

Actions: Steps to be taken – put a test facility in Beaconsville. Identify an appropriate 15 hectare site for the facility. The trial will also need to have a strategy developed to support the task of bringing industrial solar power to Beaconsville.

Following on: Beaconsville has been selected as one of five Australian solar cities, with $A5million in funding.

c. Community social change: a walking school bus

Ideals: Enhanced health of children and adults; sustainable transport; and road safety.

Facts: A multitude of players: The Walking School Bus Reference Group, The Heart Foundation, State Health organizations, Beaconsville City Council, appropriate children's organizations, local police, schools, community volunteers. Resources: Department of Health, Department of Education and Training, Environmental Protection Agency, Department of Main Roads; City Council; and the Heart Foundation. Timeframe: Immediate.

Ideas: Establish a Walking School Bus project in which each school rosters adults from each school catchment to pick up children along the way and walk them to school.

Actions: A pilot program that influences children's and families' awareness of travel behaviour, reduces the number of cars on the roads; and encourages the development of strong, safe, friendly and supportive community action. The coordinator of a local art gallery took up the challenge of developing the project, using her expertise in lobbying and marketing.

Following on: This particular project lapsed because of the heavy commitments of the proposer. However, the concept had been introduced to the city, and other schools spontaneously took up the idea.

Following on:
Third turn of the collective learning spiral

Context: The very success of the second turn of the collective learning spiral left the sponsoring local government department with a heavy overload of work. As well as strengthening the existing positive relationships with the agencies and communities which would support the projects, the department needed to renegotiate their services role to include policy and planning. At the same time the scope of the service delivery had widened from environmental services to the much wider brief of sustainability services.

The CEO decided that one solution was to train the staff of the department to use the collective spiral as a common framework for the next set of transformational change initiatives. The staff took part in a collective learning cycle to develop a strategic plan for their own services and policy.

Setting the scene: The staff numbered 15, few enough to remain in one circle. This had strengths and weaknesses. It meant everyone could share in all the learning. It also meant that the opportunity for broader discussion with clients was lost.

The training program involved a collective learning workshop followed by one-to-one discussions with the staff members on how to best fit into the team, and how best to run their first collective learning workshop.

Focus question:
How can we best form a team which delivers coordinated planning and service delivery in a sustainable city?

The session was held in a small room in the department's offices thus breaking one cardinal rule: to take the workshop participants away from their usual environments into pleasant surroundings. Another difficulty common to all team training was a limited scope to work with diverse interests. Thus there were two mistakes: to limit the workshop to office staff and not include their clients, and to remain within the office. The mistakes were reflected in a slow start to the workshop.

Ideals: A shared understanding of each participant in the team; develop skills that would allow a high level of expertise in collective learning; a collective understanding of integrated sustainable services.

Facts:

+ve: A keen hard-working team with a passion for the goal of sustainability for the city

-ve: Overwork, difficulty in knowing where one's own work fitted into the team effort.

Ideas: It can be difficult for people who work together to generate fresh ideas. Two tools set the scene for creative thinking. Having established ideals and facts, when the group met after lunch members were asked to tell each other something surprising about themselves. Some fascinating stories emerged: I was on the Berlin wall when it came down; I used to be a powder monkey in a demolition squad; I walked from Land's End to John o' Groats (the length of Britain). This was fun and allowed each person to realize that there was unrealized potential even in their small working group.

Each participant was then given a sheet of stiff white paper and coloured pens and asked to draw a picture of their contribution to the team. Each in turn then explained their drawing to the group. The group was then asked to put the pictures in some agreed working order and draw and label links between them. The discussion of where to put each picture and what to name the links opened up ideas for how to restructure the services.

Action: The CEO, who was a participant, was then asked to review the ideas and report back on their feasibility. Each participant agreed to run a collective learning process in their own area of action.

Follow-up:
Fourth turn of the collective learning spiral.

The staff members each ran their own workshops with co-learners from their own field. In the three years that followed Integrated Sustainability Services staff ran over 40 such workshops. Each workshop generated an ongoing project of its own. The full program was described as a series of systems, with their own learning cycles:

Nine Systems of Collective Thinking and Action:

System 1. Water Cycle – Creek to Coral transect

System 2. Smart City, Solar City – a new paradigm for Australia

System 3. Network Demand Management

System 4. Beaconsville Smart Grid project

System 5. Beaconsville Energy Sense

System 6. Centre of Excellence in Tropical Design

System 7. Sustainable businesses (products and services)

System 8. Carbon Community Cluster

System 9. Sustainable Beaconsville LTD.

13 Introducing new ideas: A cocktail party

Source: Art of Moving (2006)

Even a small addition can make a difference

Summary: The case studies in this chapter demonstrate the use of a collective learning cycle to enhance shared learning in an already established process in which time and resources are limited.

Background: Often there is no opportunity to mount a full collective learning spiral, as described in Case Studies 1 and 2. For contributions to conferences and strategic planning, the outcome of one collective learning cycle may be all that events allow. The issue becomes how to insert the collective learning offering into a larger process in a way that it can make a difference.

The comparison here is with a cocktail party. When time and resources are limited, there can still be a celebration, although it may have to be compressed into a short event. The two examples in this chapter are, first, a community contribution to an expert conference on changes required by climate change; and second, membership contributions to a community organization's strategic plan.

Introducing New Ideas 115

Both case studies in this chapter were hosted by the same team from a long-standing community forum. Each of the two change projects called for very different skills, however. The conference contribution called for recruiting participants from a wide range of community networks. The strategic plan called for diplomatic negotiations among members of the same organization.

Time for a single turn of the collective learning spiral was all that was available in either case. The effectiveness of the workshop then depended, first, on the capacity to draw collective interests into a common message for action within a short timeframe, and second, on the strategy for inserting that message into the main event.

Case Study 3
New ideas for an expert conference

Conference: Energy Futures

Story: Community contribution to a conference of experts on energy futures.

Transformation: From an expert monopoly to include a collective understanding.

Collective: Community members from environmental and social services

Learning: One turn of the collective learning cycle in a public form

Leadership: A human ecologist, an ecologist, a change management consultant and an educator.

Context: Energy Futures was a three day conference organized by a branch of an international organization and the regional government. Speakers from universities and government were invited to help construct a roadmap to a low energy future for the region in the face of climate change. The initial program of the three day regional conference on climate change contained no community input in its conference program.

A long-established community action group working towards 'A healthy planet for healthy people' asked the conference organizers whether they would accept a community contribution. The organizers agreed to include the outcome of a public collective learning workshop in the recommendations of the conference.

The challenge was to have a valid representation of the whole community (key individuals, community groups, relevant experts, government and non-government organizations and creative thinkers) and gather their ideas in only two hours.

To produce something useful in the time, the cycle was reduced to ideals and ideas. One short cut was to summarize the documents issued for the main conference as 'the facts'. Another was to accept that, since the conference speakers were specialists and government agencies, the workshop would invite key individuals, community organizations and creative thinkers to make up the full set of decision-makers. The action was the inclusion of the community's ideas in the conference roadmap. Another short cut was to omit the step of groups sharing their work with each other. The hosts collated the report sheets overnight for presentation at the conference summing-up the following day.

Setting the scene: The hosts emailed invitations to 50 community organizations throughout the region. Interest was expressed by 30 organizations. In the event, fifty people from twelve environmental and three educational organizations accepted the invitation to a two-hour workshop in the same building as the main conference. The time was set at 4.30–6.30pm so that people could come after school and the end of a working day.

Focus question:
What are the community ideas for a road map for a sustainable future for the region?

Five round tables of ten were set out with blank A3 sheets headed Ideals?, Facts?, Ideas?, and Actions?. Forum members roved round the groups facilitating round-table dialogue and completion of the question sheets.

Introductory speakers, the Commissioner for the Environment and Sustainability; Director of the Conservation Council, and Director of local government's Sustainability Projects and Programs, gave their ideals for action on climate change. The speakers identified sustainability as a multi-level issue:

- A global issue: the challenge of working toward human wellbeing in a healthy sustainable biosphere, with the highest national consumption rate
- A local issue: adjusting a negative carbon economy within a thriving urban economy
- A personal issue: learning to behave as global citizens working towards a sustainable future

Ideals: Workshop participants' ideals were summarized as:
- Make our city a healthy, fair and equitable democracy
- Focus on having healthy people living on a healthy planet
- Construct a coherent alternative narrative which fills us with hope and courage
- A sustainable adaptable system of collaborative management
- Reduce silos – work cooperatively across community and government
- Resolve and commit to making our city a sustainable city

Facts: The workshop drew on briefing documents on climate change and the city, prepared for the main conference. Participants spent ten minutes selecting the key items important to them.

Ideas: Different levels of energy were generated from different tables, emphasizing the power of a group to set its own identity. In this informal workshop there was no advance opportunity to balance the tables for the diversity that generates the energy. Three of the five tables of seven had high levels of energy, one medium and one lost concentration.

The ideas were collected under the following headings

Solar power
- The Solar Capital A Sustainable City
- Buildings positioned for solar orientation
- Tax incentives for changing to solar power

Transport
- Buses: yearly ticket, small buses, shuttles
- Safe and connected cycle paths
- Flexible working approaches

Finance policy
Structural adjustments in the personal and government budgets, prioritizing:
- energy efficiency
- consumption taxes
- water efficiency (grey water, restrictions, solar)

Social change (combined community, experts, government, business)
- ACT becomes an example of a negative carbon economy
- Resource re-use of all city material flows: water, food, construction, etc.
- Changed car culture, changed consumption culture
- Mandate sustainable buildings and planning regulations
- Retrofit housing, buildings, parks, shopping centres for sustainability
- Agreed community-wide SMART sustainability targets

Citizens of Canberra
- Build on where people want to do the right thing: sustainability heroes
- Make sustainability fun: forums, fêtes, fairs, learning circles, work-groups
- Encourage suburbs to be sustainability centres in themselves
- Access to learning strategies for transition: share resources, skills, ideas
- Achieve 'out there' goals by regulation if necessary
- Find best local practices around the world
- Ensure that the wealthy state that we are delivers on actions

Actions: The collated results above were worked into a PowerPoint presentation and a handout for the main conference. The time allotted to present the community findings was at the close of the conference when the expert roadmap was being prepared. In the event, the conference was running late, and the community contribution was dropped, a not uncommon occurrence for community contributions.

Following on: The host organization, the Nature and Society Forum, arranged to have the community roadmap included in the conference reports. The outcomes of the collective learning workshop were circulated to all the interested organizations. Many of the organizations reported using the list of ideas in their own planning processes.

While far from an ideal outcome, the workshop contribution had several positive consequences:
- Thirty city organizations were made aware of climate change issues
- Fifty city residents heard from the local administrators of the issues
- Fifty citizens' voices were heard, even if sadly not by the mainstream conference

Case Study 4
New ideas for a strategic plan

Project: Strategy for Healthy People on a Healthy Planet

Story: A long-standing community group collecting new ideas to update its strategic plan

Transformation: From the goal of information-sharing to taking collective action

Collective: Membership of the host community

Learning: One short turn of the collective learning cycle towards a strategic plan

Leadership: A teacher, a biologist and a human ecologist

Context: The activities of the host community group included the production of user-friendly updates on environmental issues, holding expert seminars for the community, and advocacy for environment protection through a wide range of organizations. The long-standing community group was founded in 1984, one of the earliest community initiatives to link social and environmental change.

The society had developed its aims and mission when there was still a general belief that spreading information about the seriousness of environmental degradation would alert everybody in times of change. Three decades later, there was general recognition that just spreading information is not enough to bring about change. A collective learning workshop was proposed to the organization's Board members, who agreed that it was time for a review.

Setting the scene: The workshop was advertised to all members as:

'An interactive workshop on ways of generating collective thinking and action on sustainability issues. Workshop participants will be invited to link ideals, facts, ideas, and actions in the design of a sustainability action strategy'.

Organization members were asked to help develop the focus question before the workshop. The agreed focus question was:

How can we best make use of our organization's resources to foster change towards a sustainable future?

Ideals: What should be?

- There should be a culture of awareness of ecological sustainability in everything we do
- We should draw everyone in, with more integration between special interests and forum skills
- We should all try to understand the situation as it is, and each other
- We should have a community in which new things should happen from collaboration
- We should have equality in diversity, and a fair sharing of all resources
- We should respect and explore possibilities, and listen to each other
- We should decide HOW to do things for a sustainable future, not whether to do them

Facts: What is?

+ve: A groundswell of individuals and organizations oriented to sustainability including: this organization, National Sustainability Initiative, Human Ecology program, Cooperatives; Farmers' markets; community gardens, Sustainable Schools, ActWISE, climate-friendly designs, the Village Building Co; solar technology etc etc. Government and many community members already reducing water consumption, and re-examining energy use. Ability to communicate ideas through the internet, emails, social networking; and TV programs.

-ve: A lack of public awareness of current ESD capacity, and of political knowledge of science. A gap between those who think problems are too large and unsolvable and those with well-thought out, viable solutions; between long term and short term visions; between people who don't care and the people who do; city with 70% landscapes and open spaces that need to be managed and protected; narrow vision in planning authorities, government and research centres.

Ideas: What could be? A fertile and imaginative brainstorm

- Mechanisms for developing a culture of awareness of ecological sustainability in workplaces, through incentives, witty art and cartoons like 'Art of Moving'
- Exposure of the drivers in society for unsustainable consumption

- Use IT tools: open website; wikis, blogs, one stop shop, ads for sustainability
- Revive NO Waste campaign, and using waste as resources. Ban plastic bags.
- Promote eating local produce
- Every house, new and retrofitted have its own electricity generation
- Efficient public transport system, safe nodal points, light rail, cut down overseas trips
- ANB/ANSI to be a meeting and communication hub
- Make sustainability fun: fairs, events, collaborative spaces, celebrate what is unique to ACT
- Establish the region as a Biosphere Reserve, and a Transition Town
- A Sustainability Committee (like an ethics committee for research approval)
- A 'child impact' study
- Pay the real costs and show the monetary savings of more sustainable approach
- Utilize closed school buildings for actions
- The committed communicate more with the alienated and unconverted.

What can be, in practice, now?
What can we do to achieve these ideas?

Action:

Community action:

- Live and buy locally, and collaborate with local organizations
- Urge all sectors and associations to design and maintain climate friendly homes
- Gathering evidence/examples that sustainability can be profitable
- Contact 'Art of Moving' people to design cartoons for workplaces
- Ask ActWISE project to develop a guide to local produce
- Weekly 'who-is-doing what' column in papers, and a greening tips column
- Ask Uni of Third Age to sponsor a transformational change group.

Advocacy:

- Lobby for a sustainability 'Ethics committee' for development applications, research, change of land use, and business approvals/licences
- Lobby for a sustainability committee in each workplace, similar to OHS function, called the 'Green team', with incentives, and demonstration workplace projects
- Explore Biosphere Reserve with Commonwealth government, Foreign Affairs and UN
- Lobby the local Council to oblige sellers to disclose house running costs

Own organization:

- Biosensitive Futures set up 'City Conversations' - bring people together across disciplines and from different groups (industry, government, local community, individuals, the arts)
- Establish a media contact/liaison officer
- Support Transition Towns in this city
- Work with other organizations to combine to run fun sustainability events and activities.

Following on: The flood of ideas and possible actions was not surprising given the committed, long-standing membership. The report of the workshop was delivered to a meeting of the Board of the organization. Here it was met with two sets of responses. One was delight at the range and creativity for the future. The other was more caution about implementing the changes.

The first group of members proposed a review of the organization's current direction to include the main points of the collective learning workshop. In particular they wished to include in their strategic plan the three directions recommended in the report: support for like-minded organizations, public advocacy and outreach projects.

The other response was based in concern about the move from the organization's current skills and audience. Members in this group felt that there were not enough resources to move on both directions, and so feared that the change would mean that the existing program would be reduced. They also felt that accurate, accessible information was the original reason for establishing the organization, and should be maintained.

In the event the solution was to continue with both.

Comment from a participant:

'The Earth Charter states that we should "Secure Earth's beauty and bounty for present and future generations", the same aim as the organization holding this workshop. The people at this workshop were and are very well informed that there are serious threats to the Earth's ecosystems. Yet even they found it difficult to decide how to put the actions to the words. However, the ripple effects of concerned people talking about how people are living more sustainably and healthily made the workshop worthwhile'.

14 Initiating long-term change: Opening night

Summary: In this chapter project hosts demonstrate the use of the collective learning cycle to initiate long-term, community-wide change.

Background: In the two case studies in this chapter, the first turn of the collective learning cycle is only a beginning to an open-ended change strategy that the initiators would leave behind. In one project, the change is to work towards a sustainable region, in the other towards a strong regional educational collective. Case Study 5 starts off a three-year project for initiating ongoing long-term change in 13 regional communities with a collective learning cycle for each region. Case Study 6 reports on the task of turning a formal project management committee into a collective learning team in 18 months.

In both cases, the focus is on across-the-region change from a standing start. In one case the change process is initiated from outside the region, in the other from inside.

Both case studies share the challenge of initiating collective learning as a transformational change tool, when the program will be carried on by other agencies in the future. Both start with a training workshop on the collective learning spiral, then the participants use the spiral themselves to start off an ongoing program. The aim in both cases was for the team to embed the collective learning spiral in the local practice and then withdraw.

Thus the best comparison for these programs is a program launch. This captures the excitement of starting something off, while not promising to be there in the future. In such cases, it is important to have a succession policy, in which the development of the program includes who will carry it on.

Case Study 5
Regional change: starting from scratch

> ### Sustaining Our Towns
> 'Sustaining our Towns' is a project to help reduce the ecological footprints of individuals, homes, businesses and communities in thirteen Council areas across South Eastern NSW. The project is co-ordinated by SERRROC in partnership with Clean Energy for Eternity, the Southern Rivers Catchment Management Authority and the thirteen SERRROC Councils.

Document from the original project

Project: Sustainability – Our Future

Story: Introducing the sustainability projects into 13 communities across a large region

Transformation: From a state policy statement to collective community action

Collective: Key individuals, community activists, government, environmental scientists and creative project team

Learning: Two turns of the collective learning cycle, then a third turn by 13 individual organizations

Leadership: Program leader, evaluator, administrator and facilitator followed by leadership from the community

Context: 'Sustainability – Our Future' is an edited account of the project 'Sustaining Our Towns' conducted in a large rural region in southern New South Wales, during 2007–2012. The project was funded by an Environmental Trust. 'Sustainability – Our Future' (SOF) is a five-year, well-funded program helping to introduce the idea of securing a sustainable future for individuals, homes, businesses and communities in thirteen Council areas across a region. Project partners include a regional organization of 13 Councils.

Setting the scene: The project was designed around three interrelated programs:
- A homes program (households / neighbourhoods)
- A business program (individual businesses / town centres)
- Regional development program (general communications, website / Council and local leader capacity building / regional networking and collaboration)

The staff of four covered 13 Council areas, and so were spread pretty thin. Because of this they decided to take a selective location-based approach to the project. The choice rested on a community's potential for harnessing change.

In each location the focus was on delivering both the home and business components of the program. To cater for diversity between communities 'toolboxes' of options were offered to each community rather than a one-size-fits-all approach. The SOF team were keen for communities to be choosing (within the limits of what they had to offer) what they wanted to focus on and do in their locations. Ideally this includes a mix of individual and collective activities. The project team decided to approach and engage with their selected communities through a collective learning framework.

First turn of the collective learning spiral: a training workshop

Setting the scene: The scene having been set by team dialogue on the general direction of the SOF Project, the team decided to test the collective learning spiral for its suitability as the framework with which they would approach communities. A one-day workshop was led by a facilitator familiar with the collective learning process.

Focus questions:

How to apply the collective learning spiral to community-based decision-making processes involved in our project?

How to articulate the WHAT and HOW of our project to people in a succinct way that reflects the point of difference of the project (i.e. a responsive community-based initiative)?

Outcome of introductory workshop:

Ideals: The team involved in the 'Sustainability – Our Future' (SOF) project held a collective belief that this project can contribute in its own small way to achieving real change across the region.

Facts: As in project outline on previous page.

Ideas from the training workshop:
- A useful precursor to conducting the first community meeting
- The focus questions as a framework for discussion specific to the aims of the SOF project
- Refining and rehearsing how to articulate the aims of the project
- Guidance and confidence for the initial round of community meetings

Action: The team unanimously agreed to use the collective learning cycle as the means of introducing the project to each of the 13 Council areas.

Second turn of the collective learning spiral: test run community

The SOF staff team report on their first use of the collective learning cycle in a project community.

Setting the scene:

Community meeting in a small town: Evening meeting at 7pm

Meeting aims as given by SOF project team to participants:

- To provide a forum where the community could identify for the project team how the SOF project can best support the community to become more sustainable.
- To use a community collective learning process which is divided into three parts. Participants were asked to brainstorm ideals, identify supporting and inhibiting factors for the ideals and identify ideas for action in response to the focus question.

How do we (the community and the SOF project team) work together to engage the average person on the street to live more sustainably?

1. Ideals

- Real life scenarios inspiring people to live in energy-efficient homes
- Healthy communities where people are engaged at all levels
- A walking community with government planning for this
- Total local health care with home grown food from local agriculture
- SOF project team spending time with key community people
- Inclusiveness, social justice, a social city, holistic dialogue with the community
- A project that is fun and generous
- A well-educated community that is ready to embrace change

2. Facts: Supporting and Inhibiting Factors

Supporting factors (+ve)

- Environmental targets set by the community
- Initiatives in the community already
- Subsidies available for sustainability initiatives
- Project addresses visible problems
- A community information site already exists
- Passionate people are involved
- People realizing we are in a predicament
- Local Council has environment staff
- Uniqueness of the community

Inhibiting factors (-ve)

- Bad publicity from recent programs
- Partial subsidies only
- Same people are involved (heavy workload)
- Extra funding and other resources needed
- Difficulties with current planning
- Codes restrict zoning, applying old red tape
- Isolationist biases
- Lack of local knowledge of the past
- Need tools and knowledge

3. Ideas and actions for responding to the focus question were combined

4. Actions by the community

The project coordinator asked the participants to brainstorm possible actions that could be undertaken to move their community towards ideals, taking into account supporting and inhibiting factors existing in the community. The following ideas were put forward, under the headings of:

What? How? Who?

1. What? Community garden/educational centre will help people to grow their own food, can build on existing gardens. **How?** Meeting at existing garden and determine what is needed - Council has the land. **Who?** Group of 8 people needed to approach Council

2. What? Green building sessions for builders and tradies. Enables remodelling and building of homes to include energy efficiency and water saving

3. What? Support local events. **How?** Friday Market - walkable - support of local partners and producers - funds could be leveraged by partners/sponsors. **Who?** Council parks staff, Indigenous people, market sponsors, slow food movement, police, existing businesses (e.g. Benedict House), local farmers

4. What? Education and networking. **How?** Use avenues for networking e.g. social networking sites for the project. **Who?** Engage whole community through a community calendar of events (single portal easy for people who don't have websites), set up website such as a multidimensional site (e.g. ning.com)

5. What? Support Sustainable House Day, an existing project. Make this a regular event. **How?** Funding, volunteers, sustainable house owners to open homes. **Who?** A group of sustainable homes for others to see, a range of houses modelling different types of retrofits.

Following on: The community response was enthusiastic (see poem in Box 14.1). A practical set of activities emerged spontaneously, making it a challenge for the SOF team to choose which one to fund. The decision was to search for ways to put them all into practice.

In Part 2, Chapter 11 we suggested that the success of a collective learning session is like the success of a party: the guests enjoyed themselves and wanted to meet again.

Participants' comments at the end of the SOF meeting:

'Theory/vision and practical suggestions were blended really well – a hope-filled evening.'

'When will follow up meeting be held? Should be within a couple of months and community needs clear direction about what the next step will be'.

'A process is needed whereby further development of the ideas from the meeting can be facilitated. Suggestions were a social networking group or a group facilitated by a community member or Council staffer that picks up where this process leaves off. This is essential for getting momentum to continue'.

Team reflections before and after SOF first community meeting (of 13 towns):

Before the meeting

Team discussions prior to the meeting identified particular aspects of the collective learning spiral process that needed to be changed:

- Better explanation of 'What should be?' stage of the process
- Clarification of the individual and group components of the process with steps one ('What should be?') and two ('What is?') being completed by participants individually, and steps three ('What can be?') and four ('What could be?') being done as groups
- Identify differences between the usual conflict/problem-based contexts and the Sustainability – Our Future project context which focuses on identifying and supporting existing actions
- It was decided to combine steps three 'What can be?' and four 'What could be?' into one stage that asked participants to identify actions that could be undertaken by the community and supported by the project.

After the meeting

Overall, project team members appraisal was supported by feedback from meeting participants, as follows:

- The collective learning process was effective with the variety of facilitators (different project team members facilitated each stage of the process)
- Timing and overall running of the meeting reflected the high level of preparation and planning, although a smaller venue would have been better
- Information such as funding allocations and timeframes for each Local Government Area will be included in future presentations
- The focus question needs to be succinct and linked to scene-setting PowerPoint presentation, and also linked into all stages of the collective learning cycle.

> **Box 14.1 A poem written after the community meeting held on a cold night in a small church hall in a small town**
>
> Last year
> On dark arriving,
> a doorway of light
> fell gently out
> onto the step
> A murmuring within
> that hall
> drew us in
>
> It was a
> Beginning
>
> From all our separate
> tired long days
> ideas burst forth
> We asked
> How is it
> power humbly puts down its roots?
> Change finds its wings?
> And empathy unfolds
> and lives among its neighbours?
>
> At first we struggled
> to order these thoughts
> and find respect
>
> Then as uncurling fronds of knowing-softened our edges
> we discovered
> which way was up
> and took our time
> to get there.
>
> Then we looked amongst ourselves
> and found a smile
> growing.
>
> *Melinda Hillery*

SALLY AND RICHARD 8

Keeping a learning journal

Reflecting on the regional collective learning process, Richard identified the importance of capturing and sharing the lessons learnt during the project. From her recent management training, Sally suggested a daily learning journal as a mechanism for doing this. Her notes on keeping a learning journal follow.

The learning journal as a tool for collective learning

Discussing these ideas together as a team contributes to the project evaluation process, and also fosters an ethos of respect, support and cooperation within the project team. Reflective practice as a group involves:

- setting aside time at each project team meeting (every 4–6 weeks) to discuss ideas that arise from the journal writing

- discussion could focus on collective 'problem-pooling' as well as 'what-we-have-learned/achieved?' and 'where-are-we-heading-next?'

- useful questions to orient group discussion include: What are our experiences and what is their value? What are our unfavourable experiences and how can we prevent them from happening again?

- summaries of findings from the analysis of all data collected are regularly reported to the initiating group and the project participants.

Richard and Sally thought it was well worth asking team members to keep a reflective journal. However, they found keeping a journal worked best between two or three people and least well where trust still needed to be built among members of a large group. So they decided that a journal should only be a free choice, not part of the rules of the game.

Source: (Kerryn Hopkins, Sustaining our Future)

Following on:

In the two years following the initial learning cycles in each of the 13 Council sites, a wide range of imaginative projects emerged from their communities. The impetus for each project was the introductory collective learning cycle and a seeding fund of only $5000. Some of the key projects that emerged were as follows.

Free Home and garden reviews now on offer

The Homes Program now has a team of 10 home and garden sustainability advisors 'hitting the streets' undertaking home and garden sustainability reviews across the region, along the coast and up in the mountains. We have taken about 25 home and garden review bookings over Christmas and January. We expected this period to be slow. It gave us a chance to mentor our team during their first reviews, and iron out any issues before promoting the home review service more actively.

'Sustainability – Our Future' library kits

The Sustainability – Our Future project is producing a home sustainability DIY kit for placement in all Council libraries. The kit includes:

- A step by step guide to making your home more sustainable
- Useful devices (such as a power usage meter and infrared thermometer)
- Additional resources (sustainable living books and DVDs)
- Promotional posters (to put up a display of materials to promote home sustainability and the project)

Sustainability support for Councils

Sustainability – Our Future can offer the 13 Councils we are working with support to build their sustainability skills over the next year. This includes:

- Energy and water efficiency training (planned for May in a few locations and currently under development)
- Sustainability training for Council sustainability committees (or assistance in establishing a committee)
- Assistance with the energy and water efficiency of Council facilities (particularly aged care facilities, caravan parks, community centres, swimming pools)

We also looked at the feasibility of formalizing a sustainability network between Councils in this region.

Business program

The SOF Business Program is open for business. We are currently working with about 25 small to medium sized businesses across the region. We have been successful in helping many of these businesses to access energy efficiency rebates through the Energy Efficiency for Small Business program.

Local sustainability projects

In January 2011 we approved support for local sustainability projects up to the value of $10,000 per Council area in 8 of the LGAs we cover. Most projects focus on either local food production or home energy efficiency, reflecting two of the priority concerns that we have encountered across the region. Here is a summary of the projects we will be supporting so far, each in a different town:

- Council 1 – Additions to sustainable living education facilities at the local education centre
- Council 2 – Establishment of a new community garden
- Council 3 – Council home energy efficiency education program + establishment of Action Network for Sustainability
- Council 4 – Support for further development of community gardens
- Council 5 – Implementation of energy and sustainability survey / education program
- Council 6 – Implementation of pre-ski season accommodation industry energy efficiency education program
- Council 7 – Community sustainable living education program
- Council 8 – Establishment of new community garden

Final turn of the learning spiral – three years on

Context

The project always had a finish line of February 2012. Below is the final newsletter of the project. You will note that the projects are all expecting to continue without the SOF team.

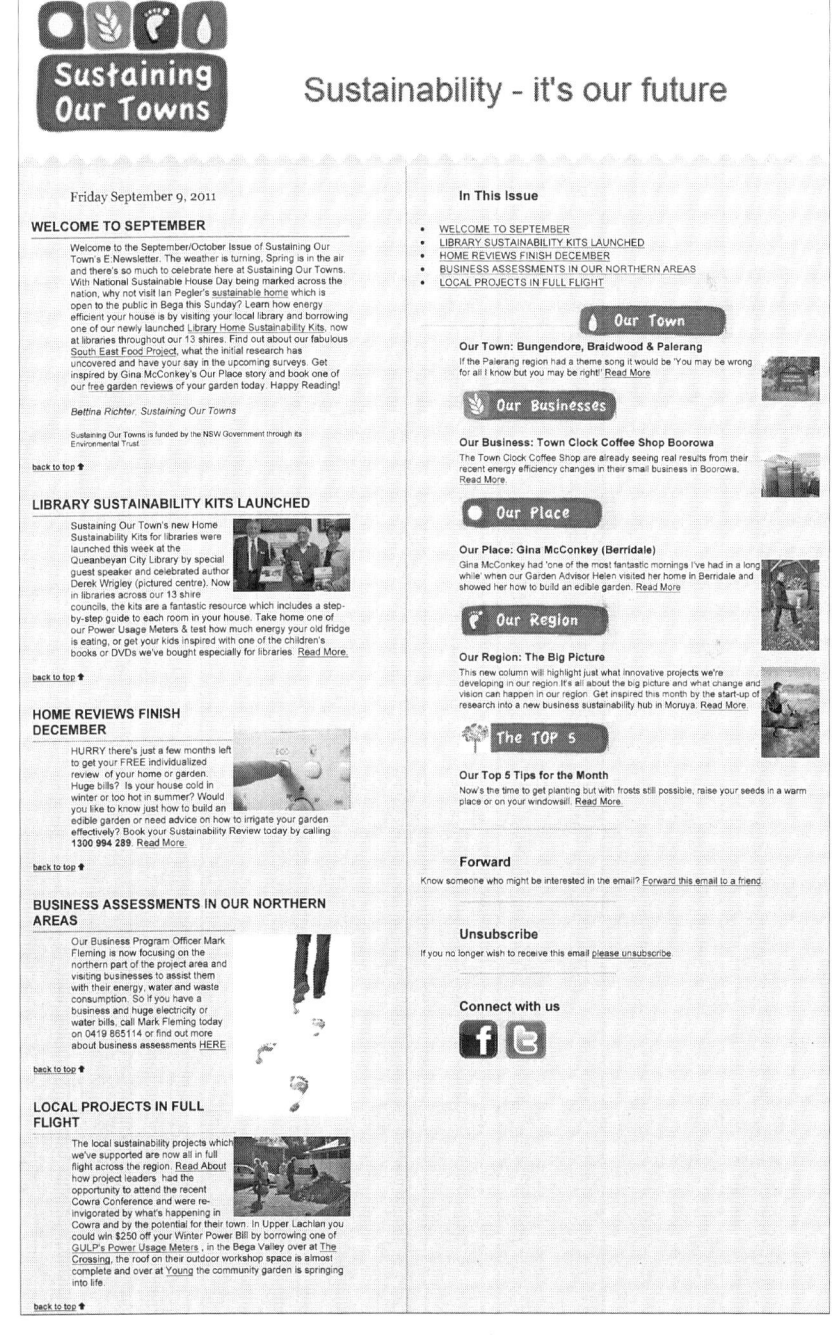

Sustaining Our Towns

Sustainability - it's our future

Friday September 9, 2011

WELCOME TO SEPTEMBER

Welcome to the September/October issue of Sustaining Our Town's E:Newsletter. The weather is turning, Spring is in the air and there's so much to celebrate here at Sustaining Our Towns. With National Sustainable House Day being marked across the nation, why not visit Ian Pegler's sustainable home which is open to the public in Bega this Sunday? Learn how energy efficient your house is by visiting your local library and borrowing one of our newly launched Library Home Sustainability Kits, now at libraries throughout our 13 shires. Find out about our fabulous South East Food Project, what the initial research has uncovered and have your say in the upcoming surveys. Get inspired by Gina McConkey's Our Place story and book one of our free garden reviews of your garden today. Happy Reading!

Bettina Richter, Sustaining Our Towns

Sustaining Our Towns is funded by the NSW Government through its Environmental Trust.

back to top ⬆

LIBRARY SUSTAINABILITY KITS LAUNCHED

Sustaining Our Town's new Home Sustainability Kits for libraries were launched this week at the Queanbeyan City Library by special guest speaker and celebrated author Derek Wrigley (pictured centre). Now in libraries across our 13 shire councils, the kits are a fantastic resource which includes a step-by-step guide to each room in your house. Take home one of our Power Usage Meters & test how much energy your old fridge is eating, or get your kids inspired with one of the children's books or DVDs we've bought especially for libraries. Read More.

back to top ⬆

HOME REVIEWS FINISH DECEMBER

HURRY there's just a few months left to get your FREE individualized review of your home or garden. Huge bills? Is your house cold in winter or too hot in summer? Would you like to know just how to build an edible garden or need advice on how to irrigate your garden effectively? Book your Sustainability Review today by calling 1300 994 289. Read More.

back to top ⬆

BUSINESS ASSESSMENTS IN OUR NORTHERN AREAS

Our Business Program Officer Mark Fleming is now focusing on the northern part of the project area and visiting businesses to assist them with their energy, water and waste consumption. So if you have a business and huge electricity or water bills, call Mark Fleming today on 0419 865114 or find out more about business assessments HERE.

back to top ⬆

LOCAL PROJECTS IN FULL FLIGHT

The local sustainability projects which we've supported are now all in full flight across the region. Read About how project leaders had the opportunity to attend the recent Cowra Conference and were re-invigorated by what's happening in Cowra and by the potential for their town. In Upper Lachlan you could win $250 off your Winter Power Bill by borrowing one of GULP's Power Usage Meters , in the Bega Valley over at The Crossing, the roof on their outdoor workshop space is almost complete and over at Young the community garden is springing into life.

back to top ⬆

In This Issue

- WELCOME TO SEPTEMBER
- LIBRARY SUSTAINABILITY KITS LAUNCHED
- HOME REVIEWS FINISH DECEMBER
- BUSINESS ASSESSMENTS IN OUR NORTHERN AREAS
- LOCAL PROJECTS IN FULL FLIGHT

Our Town

Our Town: Bungendore, Braidwood & Palerang

If the Palerang region had a theme song it would be 'You may be wrong for all I know but you may be right'! Read More

Our Businesses

Our Business: Town Clock Coffee Shop Boorowa

The Town Clock Coffee Shop are already seeing real results from their recent energy efficiency changes in their small business in Boorowa. Read More.

Our Place

Our Place: Gina McConkey (Berridale)

Gina McConkey had 'one of the most fantastic mornings I've had in a long while' when our Garden Advisor Helen visited her home in Berridale and showed her how to build an edible garden. Read More

Our Region

Our Region: The Big Picture

This new column will highlight just what innovative projects we're developing in our region. It's all about the big picture and what change and vision can happen in our region. Get inspired this month by the start-up of research into a new business sustainability hub in Moruya. Read More.

The TOP 5

Our Top 5 Tips for the Month

Now's the time to get planting but with frosts still possible, raise your seeds in a warm place or on your windowsill. Read More.

Forward

Know someone who might be interested in the email? Forward this email to a friend.

Unsubscribe

If you no longer wish to receive this email please unsubscribe.

Connect with us

Initiating Long-Term Change 137

Case Study 6
Fresh start for an existing program

Project: Integrated Urban Alliance

Story: A fragmented multi-project program re-forms into a collective social change program

Transformation: From professional silos to a strategic change team

Collective: Administrator, facilitator, researcher, community activist, funding agency, educator, program clients

Learning: Three turns of the collective learning spiral

Leadership: Funding agency and program clients

Context: An eight-partner alliance whose shared aim is to support each other to progress sustainability in the local government sector. Brought together through shared funding, the Integrated Urban Alliance (IUA) is managed through a steering committee with representatives from:

- A Shire Council and a City Council
- A research institute
- Professional Management Association
- State Environment Department
- State Local Government Department (project management)
- State Environmental Education Fund (project funding source)
- Regional Council Association

Prior to the invitation to consultants to work on team-building with the Alliance, there had been a year of formal Steering Committee and subcommittee meetings, creating a business plan and a work-plan. Multiple avenues of communication included an email network with 500 sustainability practitioners; 6 regional workshops attended by 96 officers from 53 Councils; and an information brochure and website.

A network of formal avenues of communication had thus been created to service the work of the Alliance. On the other hand, members of the Steering Committee felt that the considerable potential for collaboration among highly-skilled and experienced members of the Alliance had not been developed.

A collective learning consultancy group was invited to establish a collective learning spiral which drew on the multiple knowledges held within the Alliance.

Setting the scene:

At a committee meeting, the members of the Alliance established the project aims:

- To develop a template for collective knowledge and shared learning
- To apply the template to collective action
- To facilitate and report back on shared knowledge and capacity building
- To establish integration and capacity building over the life of the program

The task was to create a system in which all contributors to the Alliance were given full recognition, with links between them forming the basis for ongoing collective learning.

Members arrived at a focus question for the collective learning process:

How best to build a collaborative team within a formal alliance of separate organizations, each with their own project aims and processes and emphasis on different knowledge bases?

It was agreed that all the partners between them were contributing all the knowledges needed for transformational change: key individuals, representatives from local Councils, researchers, the administration and funding agencies, and the consultants introducing creative ideas. Events had proved that there was a strong tendency for each committee member partner to base their contribution on only one of the knowledges. As a formal committee this had begun to develop into conflicts. This was ironic since the purpose of the committee was to deliver an integrated alliance on urban sustainability issues.

A two-day retreat was arranged in a rural setting. Convivial dinners were arranged for the evening before and the evening during the retreat. The round tables and reporting systems were arranged as described in Case Study 2. The two days were divided into the four collective learning stages: what should be, what is, what could be and what can be.

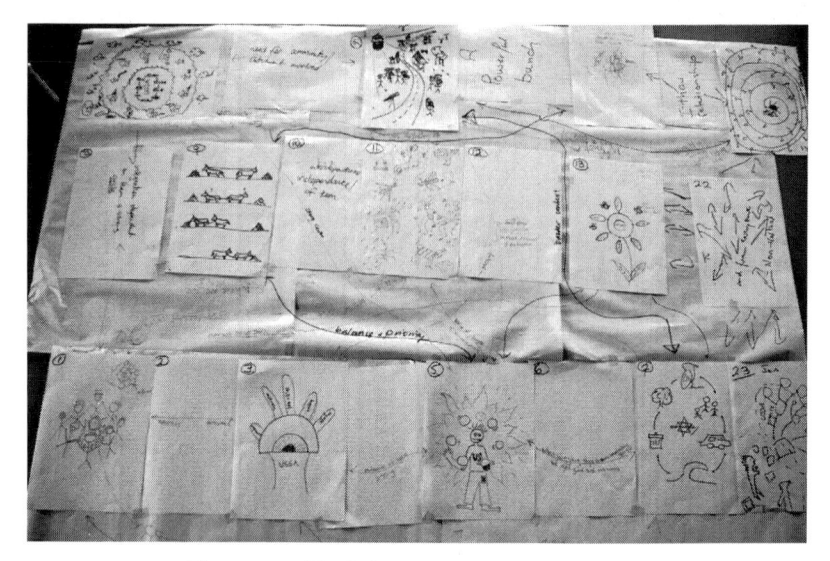

Figure 14.1 Ideals for a collective learning alliance

First turn of the collective learning cycle

Ideals: What should be? Team members were asked to draw their individual visions of their roles in the Alliance on numbered sheets with coloured pens. They then laid out their vision sheets in a square and drew a network of linking processes which turned the separate visions into a collaborative system.

For the group, the visions of collaboration were (as written on the collective drawing):

'both independence and interdependence – a strong team – a powerful bunch – a strong centre – a strong circle – balancing different priorities i.e. good juggling – when the juggling stops, the momentum starts – we become more organic and productive – we need to support each other – widen the circle to include community and catchments – broaden the context – bring together head, heart and hands' (Figure 14.1).

Facts: What is? For this stage, members identified their own preferred knowledge cultures. Note that the tools were intended to clarify the strength of each knowledge culture and the links between them. This was to set up a collaboration among the knowledges, not to enforce a hierarchy or to set priorities.

Each team member drew on their own profile for how they access each of the five knowledge cultures (see Case Study 9 for personal profile tool). People were not asked to share their personal profiles but to comment on their implications for the Alliance. Members reported recognizing their own preferred knowledge cultures, and agreed there were some that they tended to ignore. This exercise allowed Alliance members to recognize that lack of respect for each other's preferred mode of thinking was one of the reasons that they hadn't been able to form a cohesive team.

Linking activity: Questions for the group's 'On the line' tool, in the lunch break between 'identifying facts' and 'brainstorming ideas'.

Team members were asked to go outside and join up along a line. The following questions were put to the group. People crossed the line to the South for 'yes' and to the North for 'no'.

1. Who has been in the Alliance project team for more than 12 months?
2. Who lives south of the Harbour Bridge?
3. Who lives in a rural area?
4. Who comes from a science or education background?
5. Who is a bushwalker?
6. Who is a nature photographer?
7. Who has a solar hot-water system at home?
8. Who has a waste-water treatment system at home?
9. Who gets home from work before 7pm at least 3 nights per week on a regular basis?
10. Who was born before 1970?

The results showed that this was an urban group; almost all from either a science or an education background with a few from both. There were two age cohorts, early and later career; using a mix of sustainability technologies; and all working long hours. The exercise offered a greater shared understanding of the individuals and the group as a whole.

Ideas: What could be?

Team members were asked to fill in the spaces in the knowledge culture grid (Table 14.1) with their ideas for contributing to organizational alliances. The matrix identifies the contributions to collective learning each Alliance partner can contribute to the whole. The chart provided a reference sheet for each of the participating members as they worked together over the next twelve months.

Table 14.1 Grid cross-referencing multiple knowledge needs and resources

Alliance Members	Knowledges — Individuals	Community (needs and contributions)	Specialist	Organizational
Individual Members	Information about relevant program, learn and offer technical expertise	Community engagement and organizational change tools	Specialist knowledge, resources & networks	Insights into responses to sustainability from all sectors
Shire and City Councils (community)	Share experiences Rural/metro differences Mentor each other	Share experiences of engaging with local communities	Practical knowledge & experience to feed into monitoring & evaluation	Share organizational change experiences /barriers and successes
Specialist research institute	Evaluation framework update	Align research questions and methods to practical projects	Tools and resources for measuring sustainability	Access to learnings and knowledge from all partners
Funding organization	Which senior managers are progressively engaging with sustainability	Who is doing what within the Councils, and organizations	Feedback on how the Alliance was organized and what components were most effective	Bring out sustainability needs of Council members. Advocacy updates
Management Team	Understanding time constraints	Share regional experience	Pilot sharing tools, techniques/and resources	Funds to reward sustainability advances

Action: What can be? Establishing a collective community of practice among the eight Alliance members. The tools used to establish a community of practice were an affinity circle and speed dating. Using an affinity (relationship) circle (Figure 14.2a) members identified three partners within the collaboration with whom they wanted to connect. While the result (Figure 14.2b) can look chaotic to outsiders, the members know exactly who their contacts are.

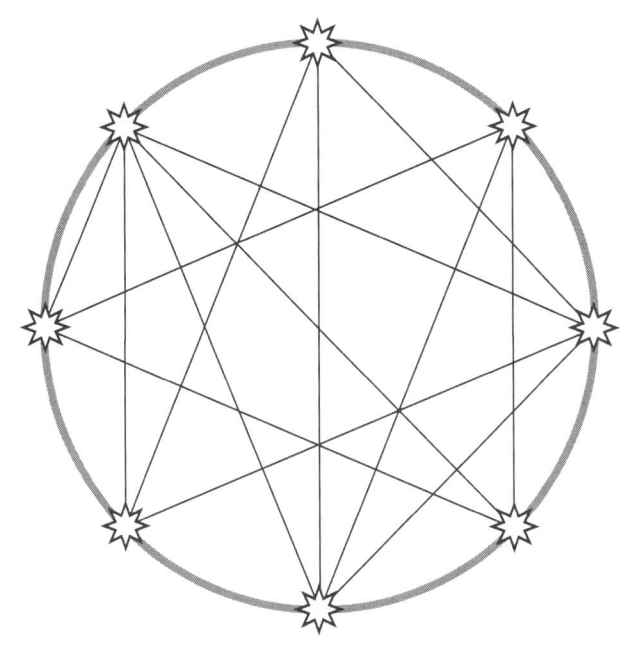

Figure 14.2a Affinity circle

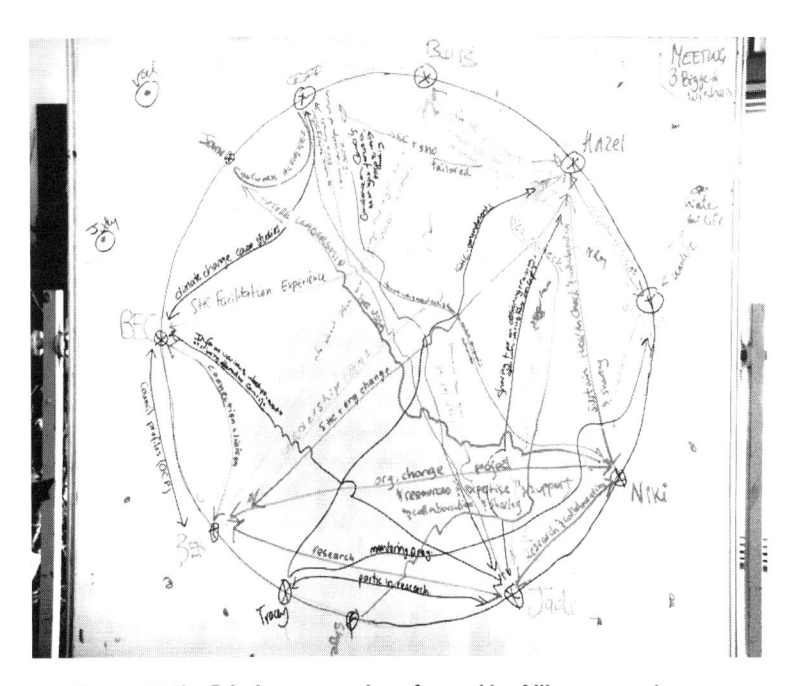

Figure 14.2b Priority connections formed by Alliance members

Initiating Long-Term Change 143

The participants then speed dated: each member spent ten minutes with each of their proposed partners to set up a collaboration which had:

- A starting date
- A timeline of mutual activity
- Intended outcomes
- Review date.

Types of responses:

Partners	Timelines	Outcomes	Review
John and Betty:	By November	Canvas regional workshops	Early next year
Paul and Peter	By Sept 15	One-page list of relevant projects	One month
Paul and Toby	By Sept 8	Clarity of research project component	Next week
Anna and Joan	By Dec	Skills exchange	Mid next year

Following on: At the start of the Alliance collective learning project, participants were asked for their personal goals. Table 14.2 lists those goals and reports on their experience at the end of the collective learning cycle.

Another type of evaluation:

At the close of the workshop participants were asked to provide a holistic focus which described the essence of IUA for them. This could be a slogan for an IUA T-shirt. These are the key ideas which taken together make for a strong collaboration and a firm alliance:

IUA is:
- a roadmap for sustainability practitioners
- collective learning for Councils
- ground-truthed pathways to sustainability
- an alliance for adaptive management
- shared experiences
- strong teamwork
- support for community drivers of organizational change
- collaborative systems of management
- sharing models, tools and learnings
- access to wide networks
- synthesis of partner strengths

Table 14.2 Sample of personal evaluations of collective learning experiences

Start of workshop goals	End-of-workshop comments
John: To gain a shared understanding of our strategic pathways, catch up on what's happening and learn from each other	Enjoyed it. Now quietly confident that we have unleashed a bit more of the project potential as a catalyst without a particular start or end time. We can support each other and this is a nice start to sustaining the process.
Paul: To lay the foundation for a smooth productive second half of the project	The 'heaviness' that was part of the project is better now than a year ago – it's a work in progress with people of passion coming together. The project admin. needs to be separated out from the intellectual elements with everybody responsible for the latter.
Joan: To learn how to better integrate, help ensure the success of the project outcomes and have fun	The enthusiasm of the Retreat is great – need to continue that momentum. The Action planning process is the key. Need administrative and process efficiency, with all team members contributing previous experience to make it easier for the core team in secretariat.
Betty: Knowledge of the common tracks of our program elements and learning	Relationships and clear roles are important and a loosening up of 'control' which was crucial at the start can now happen with increased trust between participants. This will lead to an increase in independent action by individuals who are part of the team.

Second turn of the collective learning cycle

The participants in the collective learning cycle were impressed enough to ask for a training session on how to use the cycle for themselves. This was completed at a second 2-day retreat.

Following on: The IUA project finished after three years. As part of that project a state-wide program for all the State's Councils gave professional development workshops around the State. These workshops were also based on the collective learning cycle. Thus the program launch, that is, the two initial collective learning cycles, had successfully sent off ripples throughout the region.

15 Changing problem communities: Housewarming

Summary: Two communities start with major problems (lead pollution and loss of livelihood) which require whole-of-community change.

Background: Communities can arrive at a point where they cannot see the wood for the trees, that is, they are aware of individual issues, but do not recognize an urgent need for transformational change of their whole community (first case study in this chapter); or if they do they are frozen into inaction (second case study). Citizens can become so desperate that they send out a help call to people outside their community – sometimes called whistle-blowing.

Both case studies in this chapter are of communities who had reached just such a blind alley. In one, dangerous pollution and in the other drastic social change were close to rendering their towns non-viable. In one case the collective learning team was invited in by a specialist from public health, in the other, by a group of concerned citizens. In other cases it has been the local Council that has sought help.

When the problem lies at the heart of the community, transformational change is needed to reframe the community. This can be compared with moving house, finding oneself not just with the problem solved, but living in a transformed town. Not only the new house, but new neighbours take some getting used to. Looking at a time of transformational change as moving house helps bridge the old and the new.

When a collective learning team is called in from outside a community, it is crucial that they are thoroughly briefed on the community by their host/whistle blower, backed up by their own observations and interviews, both formal and informal. In the first case study of this chapter, the call for help comes from an unhappy member of the local Community Health Centre believing that lead pollution had reached crisis levels for children.

In the second case study the collective learning team were invited in by members of a women's organization who were desperate to keep employment in the town, after the town's original reason for existence had vanished.

In both cases the towns were facing a wicked problem (see under W in Part 3), that is, a problem that lies within the society itself, and cannot be solved by existing problem-solving methods.

Case Study 7
Creating new beginnings

Project: Polluted mining town: Elizabethville

Story: High levels of pollution were causing no responsive action in the town until everyone learnt to hear each other

Transformation: From a polluted community with no future to a place where children can grow up safely

Collective: All the knowledge cultures in the town (individual, community, specialist, organizations, holistic)

Learning: Enabling collective learning among the diverse interests

Leadership: Local whistle-blower.

Context: In the early 1990s, the World Health Organization and the National Health and Australian Medical Research Council dropped the recognized risk levels from 20 to 15 micrograms of lead in a decilitre of blood for the population as a whole. The limit was 10 micrograms for children, who absorb lead at eight times the rate of adults and risked permanent intellectual impairment.

Elizabethville is a town with 50,000 inhabitants and home to the biggest lead smelter in the world. The town has considerable civic pride, like most traditional mining towns, with lots of community organizations, solid Victorian era stone buildings and a large park. In the 'safe', upwind parts of Elizabethville, the children's lead levels are below the danger level. In the downwind sectors, levels of 30 micrograms and over are found.

Treatment is difficult and painful and it can take more than 30 years to reduce to the advised level. Only 50 per cent of parents are having their children tested for lead levels or visit the Environmental Health Community Centre, including those with known lead blood levels over 10 micrograms. A recent study found that mothers of lead-affected children may deny the risk, even with quite severe symptoms in their children, because the lead pollution is an outcome of their husband's work, and they see no way of escaping or changing things.

Setting the scene: Since the community as a whole were not ready to acknowledge a serious problem, it was not possible to arrange a community-wide workshop, as in Case Studies 1 and 2. Even though the whistle-blower who recruited the collective learning team came from the local Health Centre, the Health Centre did not see it as its role to lead a response to the rising pollution.

The collective learning team identified ideals, facts, ideas and actions out of their own observations and interviews and a community workshop.

> *Focus question:*
> *How to establish safe conditions for children growing up in Elizabethville?*

Responses to rising lead pollution in a mining town

Individuals: Mother of affected child, grandfather a union official at the mine, environmental health whistle-blower.

Community: A one-industry town, employment the top priority, workers accept risk.

Specialists: Mining engineers know of no screening technology, Environmental Health Centre considers their role as science not advocacy.

Organizations: City Council members are mine employees, mine management under threat of closure.

Holistic: Security of employment the main community concern.

Ideals: Ideals of the whistle-blower from the Health Centre: A lead free Elizabethville in a mutually supportive town. One desperately concerned mother with a child recently diagnosed with a level of 20 micrograms said: 'My husband works in the mine, his father is the union organizer for the mine. I can't even mention it at home. Since our house has been notified as in the risk area for the new lead levels, it's worth nothing, so we can't move. There has to be action for the 400 children like mine'.

Facts: The lead smelter is the biggest employer in the town, followed by the health services and then the grain terminal (the surrounding country is wheat country). The town has a 12-member Council, amalgamated from surrounding small rural shire Councils. The Council has gained its first 'green' member – previously the director of Elizabethville Community Services. The Mayor is the shipping manager for the lead smelter. The smelter has been in action for more than 100 years, and has recently become part of a global consortium. Prices are not so good for lead, and the mine is under administration. The company is considering moving the mine operations to the Mexico base, where there are fewer industrial safeguards and lower wage costs. This would spell the end of the town.

Ideas: What are the possibilities for the strategic management of Elizabethville? Any expensive solution, such as compensation for mine families so they could move, or better technology that would eliminate the air-borne lead was regarded by the mine and Council as out of the question for financial reasons.

Action: Members of the collective learning team attended existing community meetings: School Parents and Friends, Rotary, Business Centres, Art Centre, Council meetings, explained their reason for being in town, and asked the meetings for suggestions for action. A groundswell of concern began to appear.

Following on: The follow-up actions of each of the knowledge cultures are as follows:

Individuals: the public airing of the issue meant that the father-in-law of the concerned parent became conscious of the emergency faced by his grandchild and other children. As a union official he was able to 'pull rank' to threaten union action.

Community: Elizabethville as a whole had been in denial. Even when there were household dogs going mad after living in the household grounds, there was no public concern. The changes in the other knowledges following community airing of the issues seemed to switch the community on to high anxiety about children's health risks.

Organizations: The Health Centre staff became politicized and lobbied Council and state governments for action; individual staff increased their community education skills.

The Council recognized the shift in the town's level of anxiety about the lead levels and that ratepayers were now willing for them to buy the polluted land: the local mine management appealed to the international management who found there was new technology which the mine could afford.

Holistic: The core concern of the town had become their children's health.

The essence of the transformation change was consciousness-raising in all the knowledge cultures, which led to the change in the core concern from livelihoods to children's futures.

Transformational change in a mining town

Individuals: Mother alerts community; grandfather alerts unions

Community: Children's health risk becomes of general concern

Specialists: Mining engineers find new technology in the USA; Health Centre embraces advocacy

Organizations: City Council buys out polluted land under the mine plume; mine management install new technology

Holistic: Children's health becomes chief community concern

The story of Elizabethville does not have an entirely happy ending. After a burst of improvement, the price of lead fell, the mine was again threatened with closure and the old fear of the citizens of losing their livelihood returned. However, the high level of risk had been reduced, and though emissions crept up again, the townspeople were more sensitized to monitoring their level of risk and advocating for further change.

Case Study 8
A fresh start

Project: Dying desert town

Story: A town faced with extinction calls on its collective resources to re-invent the town

Transformation: From facing bankruptcy to a thriving future

Collective: All the citizens of the town

Learning: Reclaiming community identity

Leadership: Lionesses, wives of Lions Club members.

Context: Rivermouth was a small town created out of the desert only forty years before. The reason for its existence was an American tracking station established as a communication resource for American troops in the Vietnam War of 1955–75. The base then continued operations to service US defence forces in general.

The US Army built an international airport, a state-of-the art global communication system and comfortable housing for their troops. They were model citizens, contributing to thriving community organizations such as Rotary, Maternal and Child health centres and recreational equipment.

In 1990 advancing technology made the Rivermouth base redundant and the American Command announced their withdrawal.

Setting the scene: Panic enveloped Rivermouth at the announcement of the US withdrawal. All employment and amenities of the town depended on the American presence. Workers became unemployed. The town's youth had no prospects. Recreation and community meeting facilities remained, but with no resources to staff them. The removal of the American planes reduced the air traffic to nearly zero and isolated the town in its desert environment.

Over the next three years, the town decreased in numbers, young people moved away, and civic organizations declined. House prices fell so it was difficult for many people to move and a small and depressed population hung on. The main income became the small trickle of tourists who had always braved the long desert drive to visit the fringing reef off-shore from the town.

As things went from bad to worse, the civil service Lions Club lost all its members. However, the companion women's club, the Lionesses remained, and became a social resource for the town. An active sub-group of the

Lionesses held a collective learning workshop among its members and some leading citizens to identify any resources that could help lift the town's morale.

*Focus question:
How can we give Rivermouth a future?*

Rivermouth: Dying town

Ideals: To return Rivermouth to its thriving past; to provide employment for the town's youth; to make the town attractive to tourists; to give Rivermouth a future.

Facts:
-ve: No employment opportunities, depressed economic base, isolation in the middle of a desert; ageing population; and a desire to return to a past which had gone forever.
+ve: Town near an untouched reef; an international airstrip; a high level of technical communication skills left in workers from the base; a small but loyal winter population of tourists; and a sense of identity in the town.

Individuals: People leaving town, low morale of those who stayed.

Community: Losing identity as an international facility; looking backward to an unrecoverable past.

Specialists: No employment, moving out to jobs elsewhere.

Organizations: Civil service organizations closing down; council with no rates income.

Holistic: Collective belief that the town has no future.

Ideas: After the initial collective meeting, informal clusters of townspeople met to build on the ideas generated at the meeting. Ideas included: using the international airport to bring in overseas tourists to the reef; providing good facilities and a welcoming town to attract a larger winter population of sun-seekers for longer stays; and to find ways of marketing the town's legacy of skills in advanced information technology.

Action: Townspeople from different interest groups pursued each of these ideas. The people who had been servicing the airport contacted colleagues in the national carrier, Qantas, and aroused interest in a publicity campaign to draw a clientele to regular flights to the town and its reef. Business from the town investigated bank funding to develop tourist accommodation near the reef.

Those with information technology experience investigated the cutting edge of information systems and found that call centres were just beginning to take hold, and they could be based anywhere.

Community members who had befriended the winter residents of the long-stay caravan parks formed a collaboration with the residents, with volunteers from both groups improving the facilities and setting up town friendship groups. And most successful of all, the town applied for, and received, a major grant from a national fund established to speed the country's entry into the digital age.

A different future dawned for Rivermouth.

Transformational change in Rivermouth: a thriving town

Individuals: People saw a future in the town, youth stayed on for new industries arriving in town, rising morale of residents.

Community: Kept identity as an international facility; maintained a high level of IT skills; looked forward rather than backward.

Specialists: Experts in tourism, nature conservation and Information Technology drawn to the town.

Organizations: Support from national airline (Qantas); national information technology development grants; wealthy council.

Holistic: Collective confidence in the future of the town.

Following on: Ten years later, the town population has doubled and doubles again every winter. Four eco-resorts have been established along the coastline. A unique global opportunity, swimming with whale sharks brings in three international and daily national flights each week. And most successful of all, the development of one of the country's first call centres arrested the outward flow of young people and even brought extra youthful expertise into the town.

Evaluation: The changes are not without their challenges. Tourist development put pressure on the delicate reef and a strong environmentalist campaign to halt all tourism put a cap on further development. The increased sun-seeker population swamped the amount of hospitality that townspeople could offer and the town lost much of its friendly reputation. Worst of all, the world-wide recognition of the need for call-centres led to reduced contracts for Rivermouth's centre.

Nevertheless, the town had not only survived, but thrived. The trigger had been the concerted effort of the whole community to build on its strengths.

16 Achieving collective thinking: Coming of age

Summary: Individuals, a lateral and a linear thinker from Western culture and a traditional Indigenous thinker, report on a self-test for collective thinking.

Background: Every individual has the capacity to tap into all the five knowledges of the Western decision-making system (see Knowledge brokering and Multiple knowledges in Part 3). By the time they are adults, most people have been socialized through pressures of schooling and employment to identify with one knowledge culture more than the others.

Added to this, Western culture has privileged specialist knowledge over personal, community, organizational and holistic knowledges. The result is that most people do not realize that they can tap into their own capacities to become collective thinkers and to make their decisions based on all the evidence.

This chapter gives three examples of how individuals can develop their own skills in collective thinking. The individuals are a lateral thinker (Dominique), a linear thinker (Janet) and an Australian Indigenous thinker (Julian). Interviews with each one demonstrate how every individual is formed by and can contribute to collective learning.

Each team member draws up their own profile for how they access each of the five knowledge cultures, and then relates that to their present activity. People are not asked to share their personal profiles but to comment on their implications for their own learning.

The basis for individual collective thinking is to recognize that each of the decision-making knowledges has its own evidence checks and its own holistic understanding of an issue.

The age for citizenship in most Western countries, the age for driving, voting and drinking alcohol, is 18. It therefore seems fitting to compare a capacity for achieving collective knowledge with an 18th birthday celebration.

Case Study 9
Lateral and linear thinking

Project: Diagnosis of individual collective thinking

Story: Individuals have preferred ways of constructing knowledge within their own cultural context.

Transformation: From a personal thinking style to a collective thinking style

Collective: Individual collective thinking

Learning: Exploration of own implicit knowledge

Leadership: Reflection by each individual

Background: All five knowledge cultures used in decision-making have developed their own sources of evidence, tests for truth and risks of ignorance (Figure 4.2). By the time they are an adult, everyone living in our era and our culture has constructed their own story by weaving together all the knowledges.

Setting the scene: These are the stories of two people seeking to become collective thinkers. They were each asked to tell their story in the light of the five decision-making knowledges. There is a challenge to you, the reader, to tell your own story in this form (Box 16.1).

Focus question:
Can I use the collective learning framework to tell my story?

Lateral thinker's story
(combining ideals, facts, ideas and actions)

Context: My story is quite long, since I am 80 years old. I have lived through the birth of aeroplanes as a normal form of transport and the internet as a normal form of communication. Some things, though, do not seem to have changed.

As an individual I have always been a risk taker, always believing that I could help things change for the better; a glass half full rather than a glass half empty person. I have started a lot of things in my time: first female research officer in the national science organization, first consumer advocate member of the National Health and Medical Research Council, Foundation Chair in Environmental Health at a national university, founder of the national Healthy Cities movement, first paper published on the relationship between environment and health, foundation of a children's theatre… and so on.

A sense of community comes from time with my extended family, with my work colleagues, in my home town, and in a rainforest or on a beach. My family of origin are conservative, my present family are radical, but we belong together. My research has been my passion and I work with other passionate people, and so I work within a warm supportive community. After being in charge of public health services in my home town I felt I knew every citizen. In a rainforest and on a wild beach I have a sense of togetherness, of being in the right place.

In our current society everyone's education lies in some specialized field. After a science degree in ecology, I spent 15 years rearing three children, a highly specialized experience. My PhD however was titled 'Holism in the University: Promise or Performance?' my first step in being a collective thinker. The next dilemma was could one become a specialist integrator, or is that a contradiction in terms? Everyone needs some sense of community in their work, and mine ranged across all the knowledges.

Relationships with organizations have always been a form of love-hate for me. Formal organizations, such as universities and employers, have a top-down ownership feel to them. Community organizations tend to be undermined by differences of opinion among the members. Yet both have the power to change things, and many skills for bringing about the change.

I can't imagine a world without the sparkle of a creative leap. The holistic sense of things coming together in a research project, of finding the essence of a hidden meaning in painting and poetry, and of feeling as one with the forest provide the reason for living.

The collective knowledge created by my learning spiral over these many years, each cycle building on the ones that went before, has created my identity, it is who I am.

Box 16.1 Building Collective Knowledge as a Lateral Thinker

Individual: Early adopter, ground-breaker, activist, costume designer, future-oriented

Communities: Extended family, work colleagues, rainforest

Specializations: Ecology, environmental management, child-rearing, public health, youth theatre, change agent

Organizations: Universities, public service, consumer organizations, environmental organizations

Holistic: Costume design, landscape design, synthesis

Collective: Identity formed from all of the above

Linear thinker's story
(combining ideals, facts, ideas and actions)

Context: Mine is a story of change and renewal. From my growing up in a working class family living on a farm, through two university courses, then several career changes, to two decades of consultancy work reaching out from a global city.

As an individual, my focus has been on seeking changes that might contribute to a more just and caring society. My bedside bookmark contains the quote 'Let me look back and to my conscience say, because of some kind act to beast or man, the world is better that I lived today'. This has been part of both my career changes and my community involvement. It probably stems from a deep sense of peace when in the more natural places in our environment, especially those that are richly remote. I would like to think that it has motivated the community involvement and the career changes that are my personal story.

My sense of community comes most strongly from my work with others who share my passion for the environment and for justice for those living in private spaces and public places where life has been less kind than it has been to me. Like-minded work colleagues, community volunteers and those who shared my interests and values while a local government representative have all contributed to my sense of community.

My career specializations have certainly not followed any traditional path. Initial training as a pharmacist was quickly superseded by a desire to contribute to human health through scientific research. During almost two decades as a research scientist, my concern for our natural world grew. It finally led to retraining in environmental management and a career switch to becoming a fulltime environmental advocate then an advisor to a national Environment Minister. Somewhere along this pathway I recognized the great importance of bringing together the science, the community passion and the policy needed to make our world a better place. That realization has seen me work in consensus-building among those who can most influence our future. That I see the perceived importance of specialized knowledge to credibility is reflected in my decision to complete tertiary training in business administration before embarking on this consensus-building role. No longer is mine a specialization viewed as reputable by any traditional discipline.

My relationships with various organizations have been intimate and at times fraught. The discord that comes with the passion of many in non-government organizations is at times extremely taxing, while the top-down inflexibility of government organizations is frustratingly disempowering. However, the discord, frustrations and passion might be the alchemy for change. During the past two decades or more I have seen the power of change when these forces come together with sound scientific or technical knowledge.

My own creative abilities are limited. Mine is a broad but linear, rather than an inventive way of thinking. My inclination is to reach a holistic understanding by working systematically through whatever I am doing. However, I value deeply the creative skills that others bring and I am keen to help bring together those different ways of knowing and to benefit from the collective wisdom and shared satisfaction that come from melding them all.

Box 16.2 A Linear Thinker as a Collective Thinker

Individual: Environmental activist seeking change through shared experiences.

Communities: Like-minded work and community colleagues, natural places especially arid and semi-arid landscapes

Specializations: Pharmacy, health, environmental science, government policy, community development.

Organizations: Scientific research, environmental non-government organizations, government (national and local), small business

Holistic: Through systematic thinking

Collective: Shared experiences with others

Case Study 10
Indigenous thinking

Project: Indigenous project manager's thinking

Story: An Indigenous thinker translates Western decision-making into their own collective understanding.

Transformation: From a Western to an Indigenous interpretation of collective knowledge.

Collective: Life, land, and language

Learning: A different form of collective learning

Leadership: The capacity to reflect on one's own understanding

Context: Working on a collaborative project with Indigenous people in their country, we tried to find a way to share how we were each making decisions on the same issue. One co-researcher was interested in the multiple knowledge framework we were working with. So we asked him if he would translate the framework into his own thinking. The result is the best translation of his thinking into Western thinking that the authors could manage.

Figure 16.1: Steve Strike photo of Uluru
in Indigenous Protected Area

Individual: The spirit of a plant or animal enters into a baby when it is born. It may be a type of wallaby, a certain sort of tree. The baby then grows up learning how their totem spirit behaves, what it needs and what their responsibilities are towards it. So the totem spirit becomes part of their individual identity and they are responsible for it for all their lives.

Community: Every community is part of, and responsible for the well-being of their ancestral land. That land cannot be in good health unless the community follows the instructions of Tjurkpa, the over-arching spirit of the land. Each community has their own Tjurkpa. The instructions come by stories, as for example when a greedy boy who eats his own totem is turned into a cockatoo.

Specialization: Every person is a specialist on their own totem. Others take their instructions on how to care for it. There are levels of specialized knowledge: knowledge known only to the elders, which cannot be shared; knowledge shared among those who live on certain lands; and knowledge that can be shared with everyone. There is also men's knowledge and a separate women's knowledge.

Organization: Every Australian Indigenous group takes their instructions for living from Tjurkpa, an ancestral spirit who knows and determines everything. Tjurkpa's instructions are so detailed and so complete that they are conveyed by songlines, long epic stories that define the landscape and the place of peoples, animals and plants in it.

Holistic: Since the parts of Indigenous knowledge are never split into separate knowledges, it could be expected that they would have no need for the integrating influence of holistic knowledge. However, Tjurkpa's instructions are so detailed that there are some core threads that weave through all the songlines. For many 'mobs' the Rainbow Serpent personifies the instructions in the form of stories.

Collective: Each person's life is as a living form that is an integral part of their natural environment. Their language is based in stories about how this living system works. Thus individual, community, specialized, organizational and holistic knowledges are never separated in the first place in Indigenous culture, and so do not need to be re-united. The arts of a culture without a written language use body painting, bark painting and dance to tell and retell all the stories.

Figure 16.2 Sally Morgan painting of Uluru

Following on: The project Transformational Change found the use of the multiple knowledges framework a useful translation tool between Indigenous and non-Indigenous members of the project team. Our Indigenous informant chose to use the collective learning cycle as the basis for his own interviews with his Indigenous community. However, he adjusted the multiple knowledges from the Western decision-making knowledges to elders and new initiates, men and women, and traditional and Western lifestyles.

17 Monitoring and evaluation: Street party

Summary: This chapter contains monitoring frameworks for a cross cultural collective learning program, and an integrated State of the Region Report.

Background: The history of project evaluation is caught up in a tangle of achievable outcomes, measurable indicators and cause-and-effect thinking. Monitoring is assumed to be a matter of making measurable observations and is honoured more in the intention than in actual practice. Further, evaluation and monitoring are often not considered at the early design phase of a project, but are plugged in at the end.

When the issues are complex, this traditional approach has the effect of narrowing the monitoring to single factors and reducing the evaluation to already determined and tangible outcomes. This does not allow for the socio-cultural changes which are features of most complex problems today.

Thus traditional monitoring and evaluation may not provide useful information for decision-making on complex problems known as wicked problems. Wicked problems are those embedded in the society that produced them, and so cannot be resolved without social change (see Wicked problems in part 3). Since any solution will bring new issues, monitoring and evaluation of the initiative have to be sensitive to the unexpected, intangible and open-ended aspects of a program or project.

Two case studies illustrate the capacity to monitor and evaluate programs that have been designed to address complex cross-cultural issues. The first case study is the monitoring and evaluation of a cross-cultural knowledge sharing program. The second monitoring and evaluation is a Regional Sustainable State of the Environment Report. The first is an emergent idea; the second has already been widely used in local Councils.

In both cases, bringing together diverse interests in a monitoring and evaluation project can be compared with a street party in which the whole neighbourhood joins in and checks each other out.

Case Study 11
Monitoring and evaluation of knowledge sharing

Project: Integrated Knowledge Management

Story: Knowledge for development has advanced from being regarded as a knowledge industry to being considered a knowledge ecology

Transformation: From separation of program delivery from its monitoring and evaluation to treating the two as concentric collective learning cycles

Collective: Decision-making knowledges of the program

Learning: Mutual learning between the project and evaluation teams from two cultures

Leadership: Research working group of Integrated Knowledge Management project

Context: A five-year grant enabled a group of international development agencies to undertake a two-way transfer of knowledge between the developed countries of Europe and the developing countries of Africa.

Recognizing that knowledge for development involves a synergy generated by all knowledge cultures working together, the questions arise: how do we achieve this synergy not only among the knowledges but also between different societies? How do we harness it for future learning and action? And how do we monitor and evaluate the changes?

Part 1 of this Guidebook described how the experiential learning cycle provides the basis for the design of a creative and dynamic knowledge initiative. The project becomes collective learning when each of the knowledge cultures are included in each of the learning stages of a shared project or program. In the case of cross-cultural learning, it is crucial that the knowledges of each culture are given equal weight (Figure 17.1).

Monitoring and evaluation of any initiative is more constructive, and of use to the participants where the design of the monitoring and evaluation matches the learning stages of the knowledge initiative itself.

Setting the scene: The participants in an evaluation of a cross-cultural program are asked to agree on a focus question. This might be;

How can we interpret the events in this initiative so both cultures can learn from the outcomes?

Then the collective learning cycle begins (Figure 17.1).

Stage 1. Ideals: The scoping of a cross-cultural initiative starts with all participants sharing their individual ideals for the outcomes of the initiative, what should be.

Stage 2. Facts: The participants establish an agreed pool of facts that set out the positive and negative parameters of the knowledge sharing initiative for both developed and developing countries: what is.

Stage 3. Ideas: Participants now go on to brainstorm the potential for change: what could be through the mutual generation of new ideas.

One of those ideas was that the knowledge sharing should be monitored and evaluated from the perspectives of the decision-making knowledges in both cultures (see Case Study 10 for an example).

Stage 4. Action: The fourth stage develops a collaborative action plan for what can be for sharing knowledge across knowledges and across countries, identifying the resources, the teamwork and the timetable. Then the same group develop a matching design for monitoring and evaluating the action plan (Figure 17.1).

This double design of program and evaluation is an ideal. When monitoring and evaluation is organized at the start of the knowledge initiative the collective learning spiral of the initiative and of the monitoring and evaluation mirror each other. However, in many cases this is not possible, since the initiative has already been planned or has already started.

In the case where the initiative has started, evidence of progress will need to be collected in hindsight. In either case, the focus questions for the monitoring and evaluation will be something like:

> *How can we best determine whether the ideals of the knowledge-sharing reflect good practice? Does the initiative meet the program's own ideals in practice? Are the facts comprehensive and sound? Are the ideas transformative and does the action plan meet the needs of the ideas?*

Stage 1. Ideally, what should be the purpose of this monitoring and evaluation? What are the ideal aims of the knowledge initiative that is being assessed?

Stage 2. Determine the 'facts' as far as the monitoring and evaluation is concerned, that is, the initiative's own ideals, facts, ideas and actions.

Stage 3. The ideas generated in the monitoring and evaluation inform the ideas of the program and vice versa, enabling mutual learning.

Stage 4. The monitoring and evaluation chooses the most appropriate tools for capturing the knowledge initiative's actions and outcomes. The results of the monitoring and evaluation are open for discussion between the evaluator and the program team.

If the knowledge initiative is addressing long term change, then the cycle is repeated at regular intervals, and the process becomes a spiral.

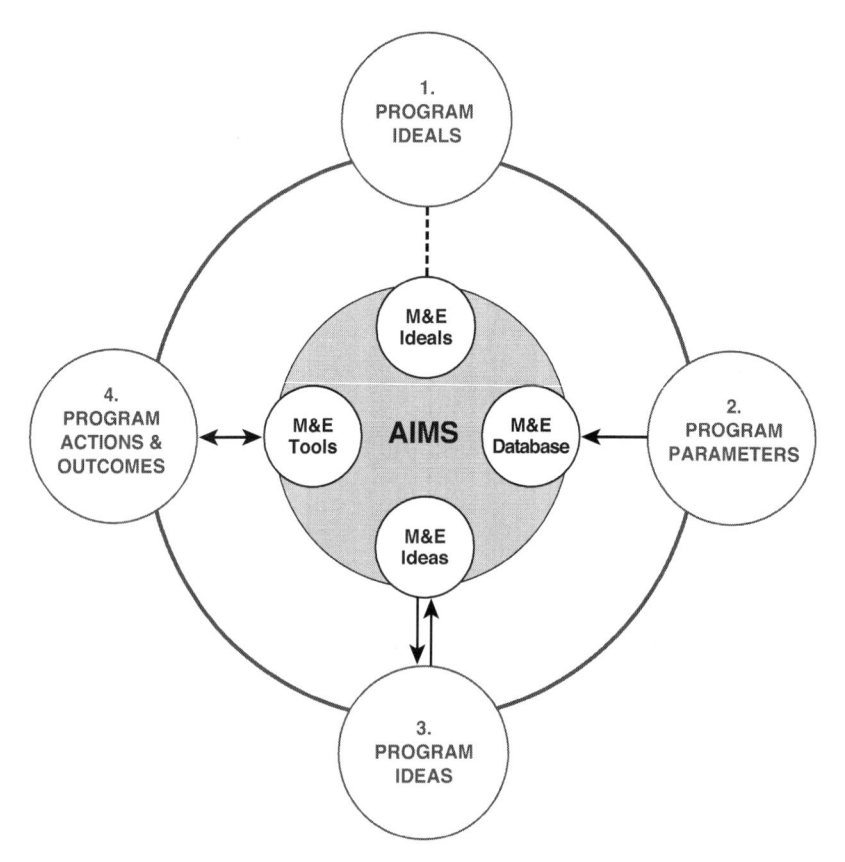

Figure 17.1 The parallel learning cycles of a collective knowledge initiative and its monitoring and evaluation (M&E)

Box 17.1 Key to Figure 17.1 in practice

Stage 1. Q. What should be? **A.** The ideals of the monitoring and evaluation (M&E) should reflect the ideals of the Program

e.g. Stage 1. of an M&E of an online learning community would take the form of a dialogue between the community and the M&E team on the ideals of the initiative and the ideals of the M&E.

Stage 2. Q. What is? **A.** The design of the Program is the database of the M&E

e.g. Stage 2. of the M&E of an interactive collective learning initiative involves documenting the learning cycle of the knowledge cultures of two countries, with its intended facts, ideas and actions.

Stage 3. Q. What could be? **A.** Exchange of new ideas between the program and the M&E

e.g. This the creative stage for each of the twin learning processes. Both the initiative and its M&E each expand their own potential through new ideas, and then share their ideas. This approach has been called illuminative inquiry since there is learning on both sides.

Stage 4. Q. What can be? **A.** The M&E selects its tools for monitoring the Program cycle

e.g. The action plan is the stage where a synthesis of ideals, facts and ideas is converted into a practical program. M&E selects the tools that are appropriate for studying each stage of the program's learning cycle.

Following on: The outcome of following the double cycle should be M&E that informs all the participants in the initiative: the individuals, the affected community, the specialist advisors, the funding organizations, and the transformative change.

In a trial of the M&E double cycle participants found that the double cycle works differently under different project management conditions. The full double cycle as in the diagram is only possible when the researcher is embedded in the project as one of the project leaders. Then the project being evaluated is designed as a collective learning cycle from the beginning.

With an external evaluator, the inner learning cycle for the evaluation remains the same. However, the outer cycle becomes a research framework through which the evaluator examines the project.

Case Study 12
Integrated sustainability reporting

Project: Sustainable State-of-the-Environment regional reporting

Story: A change in the legal reporting requirements from state-of-the-environment reporting to sustainability monitoring and evaluation

Transformation: From simple description to collective decision-making

Collective: Nine City Councils

Learning: Environment officers from each Council

Leadership: Regional Organization of Councils

Context

The search for effective tools for monitoring sustainability initiatives began in earnest in the 1990s. Examples are the Genuine Progress Indicator (developed in Australia by Clive Hamilton and in the USA by Halstead and his colleagues); triple bottom line accounting that values social, economic and environmental resources on an equal footing; and the Ecological Footprint which monitors a population's resource use.

At the time of this case study a Regional Organization of Councils commissioned a regional Sustainable State-of-the-Environment Report. The region brought together nine Local Government Authorities. The wide range of potential interests were:

- Residents, industries and community members whose day-to-day activities and experiences make up the sum of environmental health impacts
- Specialists and professional practitioners who provide expert direction
- Organizations and change agencies with responsibility for coordinating action towards the shared goal of environmental governance for health.

These interests taken together have the power to advance or inhibit the passage towards sustainability. The traditional state-of-the environment framework at the time was the Pressure–State–Response cycle in which experts on environmental issues documented the pressures on the state of the local environment from human activities and the responses to those pressures.

The Sustainability Report broke new ground in two ways. First, the Pressure–State–Response reporting framework changed to the collective learning cycle. Second, the interests of individuals and community were added to those of the specialists and the Councils.

Setting the scene

The Project team (from a monitoring centre) set up a series of monthly meetings with the Environment Officers of each of the nine Councils. The aim of the meetings was to develop an agreed set of sustainability indicators and supply matching data from each Council for each of the indicators.

The Centre also arranged a series of meetings with the interest groups of the region such as social services, youth, industries, and schools. Each meeting was asked to propose a set of indicators and how they would know whether the indicators went up or down. Both Council staff and community meetings were conducted as a collective learning cycle. Every effort was made to reach a wide range of interest groups, even to the extent of holding one cycle at 5am at the local markets to reach the market gardeners.

key to **CONTENT** of the regional SoE Report	key to **PREPARATION** of the regional SoE Report
Content: measure	**Method: identify**
Potential: Capacity to change system toward community priorities and government goals for sustainability	**Ideals** for a potentially more sustainable system
Pressures: Human activities' impacts on their immediate environment and natural surroundings	**Factors** which were most influential in achieving the sustainable systems
	Ideas for indicators for all interests to monitor the influential factors
State: 8 environmental themes	
Responses: Councils, government, industries and communities' reactions to the state of the environment	**Actions** to ensure that the indicators reach all the interests responsible for the responses

Figure 17.2 The Collective Learning Cycle as a monitoring tool

Ideal: A sustainability monitoring and evaluation system accessible to individual, community, specialists, Councils and State

Facts: Legislation identifies eight environmental themes for state-of-the-environment reports: land, air, water, biodiversity, noise, waste, Aboriginal heritage, heritage and community

Ideas: Community members identify 15 'headline' indicators for a sustainable future. Monitoring centre staff cross-referenced those to Councils' environmental themes (Figure 17.3)

Action: Commission an artist to design symbols for the indicators (Figure 17.4)

Following on: Cross-referencing the community indicators to the eight environmental themes specified in the legislation allowed collective decision-making. The next four annual sustainable state-of-the-environment reports used the indicators in Figure 17.3. The legislation then changed to state-of-the-environment reports no longer being compulsory. Six of the Councils continued the use of the community indicators.

Goal	Environmental Themes
1. NATURAL ENVIRONMENT HERITAGE Regeneration to an ideal of at least 15% of each of the original types of Regional bushland	L,A,W,B,Wa,H
2. BREATHING SPACE Everyone in Western Sydney has access to a range of safe open spaces	L,A,W,B,N
3. CELEBRATION OF WHOLE COMMUNITY: The diverse Regional community feels comfortable to celebrate itself	L,H
4. TRANSPORT AS SOCIAL INFRASTRUCTURE Shorter personal journeys to work and more accessible public transport	L,A,N
5. SUSTAINABILITY PRINCIPLES IN URBAN DESIGN Councils include ESD principles in all their planning systems	L,A,W,B,Wa,H
6. SYMBOLS OF REGIONAL SYDNEY IDENTITY Recognition of the key sites that contribute to the Region's sense of place	L,W,B,H
7. RURAL BUSINESSES VALUED Stable, transparent, rural zoning that is accepted by all the community	L,W,B,Wa
8. LIVING WATERWAYS All Regional rivers and streams available for recreational use	L,W,B,Wa,H
9. LINKED HUMAN/NATURAL SYSTEMS Wildlife corridors and urban living are mutually compatible	L,W,B,Wa,H
10. SECURE LIFE SUPPORT SYSTEMS Air, water and soil meet EPA standards 100% of days	A,WA,N,Wa
11. WASTE REDUCTION Publish amounts of waste produced and waste recycled in each Council area	L,W,B,Wa
12. FUTURE-ORIENTED PLANNING Five and 20-year projections incorporated in Council plans	L,A,W,B,N,Wa,H
13. ENVIRONMENTAL LEARNING Environmental stewardship taught in all education at all levels	L,A,W,B,Wa,H
14. ABORIGINAL HERITAGE Aboriginal communities supported in protecting their chosen heritage sites	L,W,B,H
15. LOCALIZED INDUSTRY Western Sydney industry is clean, green and local	L,B,H

Environmental Themes
L = Land   A = Air
W = Water   B = Biodiversity
N = Noise   Wa = Waste
H = Aboriginal Heritage and Non-Aboriginal Heritage

Figure 17.3 Community sustainability goals

1. Natural Environment Heritage

2. Breathing Space

3. Celebration of Whole of Community

4. Transport as Social Infrastructure

5. Sustainability Principles in Urban Design

6. Symbols of Regional Identity

7. Rural Business Valued

8. Living Waterways

9. Linked Human/Natural Systems

10. Secure Life Support Systems

11. Waste Reduction

12. Future-Oriented Planning

13. Environmental Learning

14. Aboriginal Heritage

15. Localized Industry

Figure 17.4 Symbols for community sustainability goals

18 Teamwork: Bring a plate

Summary: In this chapter, the collective learning cycle is applied to two different approaches to collective authorship of a book.

Background: There is a long-standing Australian style of party-giving which is least trouble and maximum teamwork. Each guest brings a plate of food and the party turns into a rich banquet with minimum effort.

Collective authorship can be compared to a 'Bring a plate' dinner party. With team authorship each co-author contributes a section in their own way, and then the authors meet to meld their work into a vibrant whole. If it is successful, the authors are triumphant and readers can join in the feast.

Team authorship brings a change to current publishing practice, which is mostly divided between either single authorship or a collection of separate authors. This chapter offers two different ways of creating a book through collective authorship. One is by co-writing the individual papers into a single work; the other is by compiling the book from separate papers edited in a consistent style.

Both books require a synthesizing framework if they are to form a cohesive whole. In both the cases described here the connecting framework is achieved through the collective learning spiral.

In the first example, authors and editors worked within the collective learning cycle to act collectively in producing a transdisciplinary book (Brown et al., *Sustainability and Health* 2005, see Case Study 13).

In the second example (Case Study 14) authors contributed individual papers which were then co-edited, with contributors critiquing each other's work to a collective learning framework provided by the editors (Keen et al., *Social Learning and Environmental Management* (2005)).

Case Study 13
Transdisciplinary text book, co-authored

Project: Textbook: *Sustainability and Health*
(Brown et al. 2005)

Story: Combining sustainability and health in public health practice

Transformation: From a standard specialist text to a guide to transdisciplinary future-oriented professional practice

Collective: Steering committee, expert panel, authors, and practitioners

Learning: Long lead time then one turn of the collective learning spiral

Leadership: 5 editors from human ecology, health promotion, education, sociology and public health

Figure 18.1 Book cover

Context: Public health, well-meaning as it is, has nevertheless been responsible for many of the issues facing the future in the 21st Century. As people moved from the country to the town in the 19th century, overcrowding led to epidemics of scarlet fever, cholera, tuberculosis and the rest. Hygiene, town planning, immunization and antibiotics reduced these risks to almost nothing.

From there, local and global populations have increased exponentially, until it is the planet that is crowded. The impact of that crowding is that we are drawing on the Earth's resources faster than they can be renewed. The miracles of the technical solutions have led everyone to expect a quick fix for social problems. However, it is the reliance on technical solutions that has led to the social behaviours that are damaging the planet and creating new risks to health.

For example, reliance on cars has led to atmospheric pollution and climate change, and a global epidemic of obesity from lack of exercise. Reliance on chemical controls for food crops has led to loss of many important plant and animal species and to a range of related cancers. Reliance on industries that use land and water once available for human and agricultural use, and so to the inability to survive a drought and famine. Health and environment have always been connected, but today they are connected in a different way.

The question for the future – any future – is what are our options in responding to the evidence we have of what that future could be like. We have the skills, we only need to lift the scale of our thinking. Overcoming denial, displacement activity, and avoidance behaviours that are typical responses to projections of environmental change are the very stuff of transformational change. We know about mobilizing the community, managing change, negotiating conflict, and the need for responsible local power and control (Part 3). We have professional skills to help people work towards their own imagery of a healthier, saner future, towards a future they can exchange for the risks of the present (Part 1). Where do we start?

Setting the scene: 'Sustainability and Health' is the product of a three-year process which involved many people who gave generously of their time in co-writing a transdisciplinary book on combining sustainability and health (S&H). The project started with the germ of an idea generated by an ecologist, an environmental scientist and a health promotion practitioner.

The S&H project commenced at a point of great debate and uncertainty in the field of public health, where public health practitioners were being informed about the impact of global environmental events on the wellbeing of populations, and yet they were not included in the debate.

Three collaborative groups were essential for developing a coherent transdisciplinary text. A steering committee was made up of the S&H project team, the funding agency and potential users of the textbook. An expert advisors' group of 12 international thinkers on sustainability provided direction, giving permission to use their work extensively, critically reading various reports and participating in web conversations. A third group was the 36 universities teaching public health in Australia who supported the project.

Focus question for the book, developed by the three support groups:

What is the role for public health practitioners in global sustainability governance?

The opportunity to answer this question came from funding for public health education and research. In September 2002, 60 people drawn from the five interest groups (individuals, community, specialists, organizations and creative thinking) were invited to attend a writing workshop. The task was to prepare the first draft of a textbook on sustainability and health.

Possible participants were drawn from each of the interest groups of key writers on the challenge of linking sustainability and health, community activists on the topic, specialists on sustainability and on health, government and non-government agencies and creative thinkers. Thirty people accepted the invitation.

The S&H project team consulted with their expert advisory group and then prepared briefing notes to accompany an invitation to a collective learning workshop.

Briefing notes for members of a Sustainability and Health (S&H) workshop

1. The Sustainability and Health Project has the somewhat ambitious goal of introducing sustainable development principles into every Public Health teaching program in the country.

2. The expert advisory group, who will be at the workshop, have recommended that the Sustainability and Health Project seek to develop the capacity of Public Health teaching programs to bring a new direction to public health.

3. The preliminary literature review has identified a strong literature supporting the integration of Sustainability and Health, and we can supply you with a list of over 20 valuable websites.

4. A web-based discussion over the past six months has concluded that there are serious hurdles to be faced in incorporating sustainability themes into Public Health, e.g.:

- Lack of general professional recognition that the global environment presents a major public health issue
- Difficulty in taking environmental warnings seriously, when life expectancy has increased and child mortality decreased world-wide
- Considering environmental degradation a problem of the third world, and not the industrialized world
- Relegating sustainability and health to the specialized field of Environmental Health, whereas they are matters for all areas of public health
- The challenge of applying the precautionary principle in a field that is traditionally evidence-based.

The curriculum design will be based on an integrated decision-making framework following the collective learning spiral:

- **Ideals:** To apply the principles of sustainable development in professional practice
- **Facts:** Health issues arising from the interaction between people and place
- **Ideas:** The potential for constructive management of social and environmental change
- **Action:** To evaluate examples of emerging best practice

Attached you will find a possible book outline. If you are willing to consider participating in the further development of the manuscript would you contact the Project Manager. Our national team is available to support and advise you in adapting the materials to your particular needs.

Signed, Editors ....

First turn of the collective learning cycle

Thirty people accepted the invitation. Luckily, they included the full spread of interests. The four learning stages of the collective learning cycle were approached step by step, rather than all together in the one workshop. The results of the learning stages are included here, not so much for the actual content as to demonstrate the gradual but orderly development of a book from very different contributors.

Ideals: Using social media. To gather ideals from all the participants before they actually met in the first workshop, the S&H team made good use of social media. A blogging site was made available to all participants. This was used unevenly, so individual emails and phone calls were needed to ensure a full response. Ideals were:

- to bring transformational change to environmental governance and public health practice
- to design an effective strategic planning process for the management of change
- to develop a transdisciplinary approach to public health issues
- to give direction to the complexity and uncertainty of environmental futures

Facts: Briefing papers: Extensive briefing papers abstracted information from the ideals, facts, ideas and actions already described in the literature. These papers eventually became content materials for the book. They also generated comments on the blog which formed the basis for the eventual book outline.

Book draft outline: Sustainability and Health chapter headings

1. Living
2. Listening
3. Grounding
4. Knowing
5. Scoping
6. Acting
7. Innovating
8. Managing

Ideas: Two-day workshop. The participants arrived to a convivial dinner the night before a 2-day workshop. At the dinner, people from each knowledge culture were put into mixed chapter groups to meet each other. From this experience, groups were renegotiated the next morning. Most of the groups remained the same, but everyone had the opportunity to choose after they had met each other.

The workshop proper opened with a value line (Box 18.1). This allowed the participants to get to know each other's values and the options early in the meeting. The ideals and the facts having been canvassed before the workshop, the two days could be allocated Day 1 to Ideas and Day 2 to Actions.

Box 18.1. A Value Line

In a value line, a line is marked across the room, ending in twin poles of opposing ideas. Here the poles were 'the changing social environment is the crucial issue' and 'the changing natural environment is the crucial issue'.

Participants ranged along the full length of the line with three clusters, one close to each end and one in the middle. The discussion of why people stood where they did on the line opened up excellent dialogue both within and between the clusters.

The dialogue between the clusters at the poles established much common ground on social and natural environments, the theme of the transformation project.

To everyone else's surprise the cluster in the middle spread out around the other two clusters. The members explained that they were embracing both positions in search of a fresh synergy, rather than balancing the two.

Day 1: Group ideas. After the welcoming value line, participants went to mixed tables, each with the five knowledge cultures. From sharing among the groups, the chapter content began to be filled in. Each table of eight to nine participants presented the fruits of their discussion to the other three tables before lunch, for exchange.

Lunch included a brief walk in the nearby Botanic Gardens and return to a 'fun' exercise (the 'on the line' activity used in Case Study 6). Further development of the chapter themes in the light of the cross-group discussion by the then relaxed minds led to the following book outline:

1. **Living:** Public health and the future of life on the planet
2. **Listening:** Coordinating ideas on sustainability and health
3. **Grounding:** Coordinating contexts for sustainability and health
4. **Knowing:** Re-integrating knowledges of sustainability and health
5. **Scoping:** Sustainability potential, planning and progress
6. **Acting:** Practitioners as actors for sustainability and health
7. **Innovating:** Practitioners as change agents
8. **Managing:** Public health leadership in sustainability and health

The working day concluded with a plenary discussion and presentation of the chapter outlines from each group on one large sheet of paper (or individual whiteboards). At the informal dinner that followed, members had so much energy for the task that they kept working on their chapters.

Actions: Authorship and editing. Day 2 opened with a plenary discussion of the emerging shape of the project overall. Participants from the Day 1 transdisciplinary groups were then asked to choose which chapter they would like to work further on, so long as there were still equal numbers in each group. The new interest groups (sometimes called affinity groups because of their shared interest) worked on developing and refining their chapter content. Groups lunched together to finalize their presentations before an after-lunch plenary. The plenary identified gaps and removed overlap. The further refinement can be found in Box 18.2.

Box 18.2 Book Outline as Allocated to Author Groups

Chapter 1.
Living: Public health and the future of life on the planet
- 1.1 Why this book and who is it for?
- 1.2 Global ecological integrity
- 1.3 Knowledge and action
- 1.4 Sustainability in public health practice

Chapter 2.
Listening: Coordinating ideas on sustainability and health
- 2.1 Sustainability and science
- 2.2 Sustainability and economics
- 2.3 Sustainability and environmentalism
- 2.4 Sustainability and feminism

Chapter 3.
Grounding: Coordinating contexts for sustainability and health
- 3.1 Strategic frameworks
- 3.2 Health-based frameworks
- 3.3 Social justice frameworks
- 3.4 Ecological frameworks

Chapter 4.
Knowing: Re-integrating knowledges of sustainability and health
- 4.1 Evidence-based public health
- 4.2 Multiple knowledges
- 4.3 The need for synthesis
- 4.4 Tools for synthesis

Chapter 5.
Scoping: Sustainability potential, planning and progress
- 5.1 Decision-making
- 5.2 Pattern languages
- 5.3 Negotiation and planning
- 5.4 Evaluation strategies

> **Chapter 6.**
> **Acting: Practitioners as actors for sustainability and health**
> - 6.1 Systems thinking
> - 6.2 Planning the Action
> - 6.3 Acting individually
> - 6.4 Acting together
>
> **Chapter 7.**
> **Innovating: Practitioners as change agents**
> - 7.1 In structural change
> - 7.2 In capacity-building
> - 7.3 In education
> - 7.4 In research
>
> **Chapter 8.**
> **Managing: Public health leadership in sustainability and health**
> - 8.1 Leadership and structural change
> - 8.2 Hierarchical organizations
> - 8.3 Networked organizations

The final half-day was spent in a plenary discussion deciding who wished to continue as an author, and who as a support network. Everyone offered to stay involved in the process. Authors made arrangements to meet with fellow-authors of their particular chapter.

The workshop closed with a circle, in which everyone shared one word which described the workshop for them. Then the group formed a drumming circle, in which they rapped their knuckles on the floor, developing a shared rhythm and enhancing the sense of togetherness.

Following on: Authors for each chapter formed email groups, with a chapter editor in each group. The book editors negotiated with possible publishers and completed the final tasks of copy-editing and compiling glossary and index. The book was taken up as a set text in at least five public health programs and reviewed favourably in several journals, each of which made a comment on the cohesiveness of the book given it had so many authors.

Case Study 14
Transdisciplinary textbook, edited papers

Project: Textbook: *Social Learning and Environmental Management* (Keen et al. 2005)

Story: Developing social tools for environmental management

Transformation: From a technical to a social basis for resolving environmental problems

Collective: Community activists, environmental scientists, local government

Learning: Long lead-in then collective learning cycle

Leadership: Transdisciplinary forum, three human ecologists.

Context: Since the 1950s, environmental management has been based in environmental science. As the 20th century continued and the environmental issues from technological causes increased exponentially (think Silent Spring, Limits to Growth, Chernobyl, Bhopal) more and more people came to the conclusion that there were limits to technological solutions.

It became increasingly obvious that lack of management capacity, not lack of technology was the chief block to resolving major environmental issues. Publications that addressed the social change skills in relation to 21st century social and environmental issues began to appear. These publications became more and more transdisciplinary, that is, more and more likely to contain contributions from the affected community, specialist researchers and management.

In one of these publications, a diverse range of authors wrote chapters on problem-solving at different scales of environmental governance. The comparison is with a 'Bring a plate' party where each guest (writer) brings their separate contribution, confident that the editors will turn it into a feast. Here, each chapter was written independently and then connected within a collective learning framework.

Setting the scene: A human ecology group at a conservative university felt frustrated that, in spite of increasing environmental degradation, the pace of change in environmental governance remained slow. The group contacted other areas in which there might be similar frustration, seeking out those who might like to contribute to a collection of papers. They found that the frustration extended through many academic departments, community activists, and local government environmental managers. Three members of the discussion group agreed to take the lead in designing a transdisciplinary consultation process establishing the issues and themes that might lead to a coherent set of papers.

The design of the consultation process was based on a turn of the collective learning cycle. Stages 1 and 2 (Ideals and Facts) would be canvassed online. From that material, a two day workshop would start at Stage 3, Ideas and finish with Stage 4, Action.

Following this design, the 70 respondents were linked in an email group, and asked to nominate key issues in accepting social learning as a crucial strand of environmental management. A summary of the nominated issues follows:

- Consensus building and resolving conflict
- Policy directions and institutional frameworks
- Behavioural change
- Language, communication and shared understanding
- Research structures and paradigms

Roughly equal numbers had put forward each topic, so the respondents were grouped into email groups under these headings. Three weeks before the 2-day workshop, they were then asked to form online focus groups to discuss their answers to two questions, the first and second steps in the collective learning cycle. A summary was prepared to provide a brief overview of input from the various groups, so that all participants had some familiarity with the input from others.

Focus question:
What will indicate to you that the Social Learning for Sustainability workshop process has met your aims for the field?

Ideals: Individual ideals

- 'group identities' informed by merging individual perspectives
- promotion of critical deliberation
- open communication lines furthering development of networks
- clear individual and intra- and inter-group understanding
- personal boundaries challenged and thereby extended
- complex dialogues that cut across disciplines
- people motivated to learn new things from each other
- understanding and valuing difference, seeking to achieve openness and trust between differences

Facts: What are the +ve and -ve factors for advancing social learning and achieving progress towards sustainability?

+ve:

- 'hearing' the wisdom of a broad range of perspectives
- deliberative democracy, critical theory
- 'social learning' itself as a theoretical framework
- experience from 10 years of fieldwork
- community development, social theory, behaviour change
- critical realism, social capital, social marketing
- Kolb's experiential learning
- adaptive management and participatory action research
- field work; 'place' theories, contextual issues and beliefs
- tapping into energized and committed individuals

-ve:

- lack of Indigenous people's involvement
- cross over from biophysical to social sciences very difficult
- no one-size-fits-all framework can work
- implicit knowledge of social learning not recognized
- transdisciplinary communication problems

- agency/power blocks to individuals/community action
- institutional reform to achieve more openness and participation
- no recognition of community ownership and commitment to change
- power dynamics of integrating expert and community knowledge
- need time and space for reflection and constructive dialogue

The ideals and the positive and negative factors were compiled and circulated to the whole group in the week before the two-day workshop. This left the third and fourth stages of the learning cycle, ideas and actions, as the task of the two day workshop. Thirty people from the larger group accepted the workshop invitation, and were allocated to one of the original project themes: conflict resolution; policy; behaviour change; language, communication and shared understanding and research.

Ideas: The brainstorm of ideas came up with the core theme of the book: five inter-related stands which together make up social learning (Figure 18.2).

SOCIAL

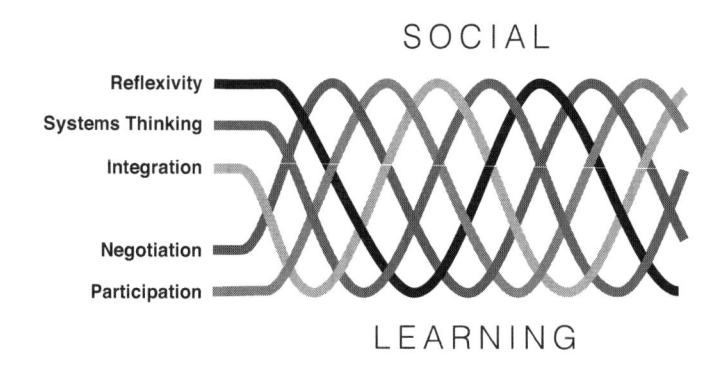

Reflexivity
Systems Thinking
Integration
Negotiation
Participation

LEARNING

Figure 18.2 The five braided strands of social learning

The basis of the eventual book design was to link the five strands of social learning to different levels of change management, theory, community practice, Council practice and individual practice (Box 18.3).

Box 18.3 Book outline

Overview
1. Frameworks for interpreting social learning and sustainability

Exploring the boundaries
2. Traditions of understanding: language, dialogue and experience
3. Complex adaptive systems: constructing mental models
4. Civil Society: bridging and bonding capital
5. Communities' self-determination: whose interests count?

Re-negotiating the boundaries
6. Linking community and government: islands and beaches
7. Combining people, place and learning: making the connections
8. Bridging scales and interests: collaborative learning
9. Science communication for scientists: reshaping a culture

Re-drawing the boundaries
10. The ethics of social engagement: learning to live and living to learn
11. The reflective practitioner: practising what we preach
12. Changing governance: embracing the precautionary principle
13. Social learning and the self: experiencing felt knowing

Re-imagining the boundaries
14. Beyond boundaries: lessons from the past, guidelines for the future

Action: Participants in the collective learning workshop were invited to contribute a chapter to the book. Twenty authors agreed, and decided to meet in a further workshop in order to coordinate their contributions and maximize cross reviewing of the chapters. This time the workshop was based on dialogue.

Editors' letter to potential authors

'You are being asked to contribute a short paper (approx. 3000-5000 words) to the writers' workshop. The draft papers will be briefly presented and discussed at the workshop as background to discussions on the key themes. Revisions will then be made based on the discussions, before the papers are used in the publications/websites. A working title for your paper and an abstract of approximately 300 words which relates your topic to one or more of the key themes and/or central questions should be included. For those unable to write a paper, we would still welcome your participation in the workshop.'

Social Learning for Sustainability: Writers' Workshop 2

Setting the scene

The writers' workshop that followed did not follow a collective learning cycle. Rather, it used the idea of critical loyalty to develop the book from a collection of separate papers to a coherent story.

Ideals: Critical loyalty describes the way in which a group can share an ideal and yet work together to interpret it in many different ways. The group remains loyal to the idea, while free to criticize each others' contributions for whether they advance the ideal. The criticisms are not addressed to the individuals, but to their capacity to help achieve the group goal. In sport this is called 'Play the ball, not the man'. Sexist as that comment is, it applies to any collective enterprise.

Ideas: Contributors to the book were grouped under its main themes, and then asked to criticize each other's paper for relevance to the book as a whole, style, coherence and accuracy.

In Table 18.1 below, the workshop design (column 1) is matched to the actions to be taken in the spirit of critical loyalty.

Table 18.1 Design of writers' workshop and book commitments

Time	Activity	Commitments
9:00 – 10:30	**Plenary** The vision for the book as an integrated whole. Overview of book's structure and content. Core integrating themes for the book came up with the core theme for the book: (discussion of Chapter 1). Discussion of connections between chapters through themes.	To ensure that the book is an integrated whole and all authors are comfortable with the substance and themes presented in the introductory chapter.
10:30 – 11:00	Morning tea	
11:00 – 12:15	**Small group discussions 1** Discussion of 3 chapters	To provide constructive feedback to authors. To increase awareness of the content of other chapters. To consider whether the selected themes in Chapter 1 are being addressed, and whether other themes are arising (ie. the 5 strands of social learning derived from the last workshop)
12:15 – 1:15	Lunch break (leading into a working session)	
1:15 – 2:30	**Small Group Session 2 (over food & drink)** Discussion of 3 chapters	To provide constructive feedback to authors. To increase awareness of the content of other chapters. To consider whether the selected themes in Chapter 1 are being addressed, and whether other themes are arising (ie. the 5 strands of social learning derived from the last workshop).
2:30 – 3:30	**Plenary** Discussion of coverage of core themes Issues for integration across book chapters. Concepts and issues of the concluding chapter. Revisiting integrative frameworks	To critically reflect on the core themes and any other emerging themes which need to be integrated across the chapters. To ensure that there is a synthesis of ideas in the concluding chapter.
3:30 – 4:00	Afternoon tea	
4:00 – 5:00	**Follow-up & Networking for Future** Time to complete discussions on integrating frameworks (if necessary) Time to pursue any issues that emerged through the day and any future activities/projects.	To allow time for synthesis and critical reflection. To allow time to creatively think about the potential for future activities supportive of a social learning network (and other initiatives floating at the last)
	Social Dinner to be organized for those wishing to carry on.	

Following on: Following the workshop, the twenty authors honoured the commitments that they made in the writers' 'Action' workshop. The book was launched six months later after considerable work by authors and editors. Reviewers commented on the value of the practical case studies that made up the book. They also recognized that treating environmental management as collective learning was a transformational change from current practice. A survey of the literature on social learning a year after publication listed the book as a major influence in new thinking about social learning.

19 Working from the Guidebook: Going it alone

Summary: This chapter presents the feedback received from two groups who have, of their own initiative and with some assistance from others experienced in the use of the process, followed the collective learning spiral to address significant change.

Background: As we go to press, a range of transformation projects have been trialling a draft of the Guidebook with no prior training. The raw results from two workshops in non-Western countries are presented here for those who wonder how the collective learning cycle goes when you start from scratch in a different context. Both examples are in progressive management regimes in non-Western cultures. The first is a strategic planning session for a school board in Bali, and the second a workshop for transformational change in health care in Ethiopia.

Case Study 15
Strategic planning for progressive education

Story: The manager of a progressive school wished to enable a school community to take the school to the next stage of development

Transformation: Establishing a progressive school in a traditional community

Collective: Teachers, committee, parents, students, employees, experts, banjar (people with a common interest) and the community

Learning: Start off with a collective learning cycle for the school community

Leadership: Consultant facilitator and school principal

> *Focus question:*
> *How can Montessori Bali future school be truly integrated, exemplary and ground breaking in its thinking, action and delivery in education, environment and society?*

Context: Reflections from the organizer of the collective learning process for Montessori Bali School, 26 November 2011.

For about a year prior to the Collective Social Learning workshop, discussions had taken place at Owner/School Board level about the possibility of moving to a new school site. Currently the school is operating on two separate sites due to lack of space at the original preschool site. The elementary school grew out of the preschool when parents did not wish their children to leave, so pushed for the next phase of the Montessori program. As enrolments increased the school leased premises about five minutes' drive away in the same general area.

Enrolments were still increasing and there was now a need for a much larger campus. How best to do this and how could this be achieved? Several meetings took place with ideas tossed around but no real action taking place. As Principal, I added my input about the difficulties of managing a school based on two separate sites and about the high demand for Montessori education in the Bali region. A new purpose built school on one campus would be a good way to move forward.

My son Nigel attended one of the meetings and offered to run a collective social learning workshop to help us envision a future school for Montessori in Bali. He had attended many CSL workshops in Australia and had taken the training and was convinced of its efficacy. Everyone thought this a great idea and so a workshop was planned for 26 November 2011.

As a Principal, I'm a little bit of a control freak and I like things to be well planned, well organized with everyone knowing what is expected and details worked out in advance. When I mentioned this to Nigel he said not to worry he'd take care of everything. All I needed to do was to confirm the date and venue. I went ahead and organized this.

At one stage Nigel asked me if I knew someone who could help him during the workshop. Someone who had the skills to help people think or imagine outside of their usual constraints. Immediately I thought of the perfect person and sent her an email asking for her help and sending her some explanatory material on the nature of the workshop. This person's motto on her business card reads:

'the power of people; more sustainable living; a better planet for everyone'.

From my previous dealings with her I thought she would be a really good person to help us with the process and she had seemed very interested in Montessori education. So I was surprised when I received her email (Box 19.1). I am including my request along with her reply. (I've removed her name).

Box 19.1 Correspondence Between Workshop Organizer and Potential Consultant

From: principal@montessoribali.com

Organizer to Consultant

HI

I'm sure I've already told you this but we are planning a new school to open in 2014 or thereabouts. We are having a social collaborative workshop facilitated by my son Nigel on November 26th at our elementary school and we need two 'coaches' to work with the breakaway groups. As I'm part of the process I cannot do this myself so I'm wondering if a) you are interested, b) available on that date. If you are interested in doing this, Nigel would need to meet you the day before (25th) to go through some details with you. It is all voluntary so there is no remuneration other than a good lunch!

I would love you to be involved in this process and I feel you will have much to contribute yourself!

What do you think?

Wilma

Consultant to Organizer

Hi Wilma and thanks for thinking of me. I'm a strategist and a creator. Please correct me if I am wrong but I think you are looking for a facilitator for this meeting. My preconceptions would impede your process not to mention my lack of patience with collaborative decision making! But thanks for thinking of me and good luck. xx

Organizer to Consultant

Hi, This is an ideas meeting to kick start the process and I thought your fresh viewpoints would enhance the process.

Never mind, I appreciate the prompt reply, we'll hopefully catch up another day.

Cheers, Wilma

Consultant to Organizer

Hi Wilma - no problem! I did read the document you sent around and I think what you are looking for is less someone with ideas and more someone who can facilitate and get everyone on board with the process. Having done this work with parents at ... school I will tell you that this is a mine field and requires someone with very good skills at keeping things on track and keeping people feeling good. By all means open it up to the community but I would do so in a much more structured setting and manage their expectations on outcomes.

Following on:
The Montessori collective learning team outcomes

From Organizer to Authors

Dear Val, I've included this correspondence because it surprised me. I thought the experienced consultant would be good for the role but her response was not what I expected. That was the first surprise.

The second surprise was my own response to relying on someone else to organize the workshop (i.e. Nigel). We have different approaches to doing things and I had difficulty not interfering and checking that he was following up on the invitations etc. (Control issues). He delegated the invitation to a graphic artist in China and we didn't actually finalize the wording of the invitation. Nigel had sent us an email with the wording but as we (as a group) were busy with other matters we didn't follow up and really look at what it said and decide if it truly said what we wanted it to say.

The response to the invitation was good although there was one parent who said he was very disappointed with the wording as he felt it was very déjà-vu. Everyone had to build for sustainability these days so what's new about this workshop?

Those invited then started to ask me what the workshop agenda was and what would be discussed. David, the head of our school board also wanted to know and I began to get nervous about the whole thing as I couldn't answer them. All I could say was, this is a first step in finding out what we want for the future of our school, please trust the process. This was hard for me to do as I didn't know if I could trust the process!!

I had several discussions in the week leading up to the workshop with parents who wanted to know more and I had to keep saying just trust the process. During this time I read your draft handbook and began to get very excited indeed at the possibilities of the day and I became more confident in my answers to parents.

Everyone enjoyed the workshop although I noticed one or two did not seem to enjoy it as much as others. The day was just a little too long and this was partly due to the fact that we waited for an important guest before starting and so started late. Some participants had to leave on time so the last session was a bit rushed.

Feedback from the workshop over the next few days was very positive and people started to come up with new ideas and discuss possible options. It was as if everyone had been given permission to be creative as a result of the workshop. Personally, I found my mind buzzing with ideas and I found that my thinking had shifted to a more open 'can do' attitude. Multiple possibilities seemed now more achievable.

The third surprise took place at the meeting. The parent who was very disappointed with the invitation and who has a very strong personality turned out to be a great listener as well as contributor! He raved on about the workshop for days afterwards and offered his time and expertise to the school.

All in all, it was a very positive experience for me and for most of the participants. I suspect we will continue to hold CSL workshops to further refine and develop aspects of our new school.

Wilma Grier, 17/01/2012

Authors to Organizer

Dear Wilma,

Thank you so much for your report on your experiences in the collective learning process. We are so happy that it all turned out so well, but perhaps not surprised. Whenever people can commit themselves to the process, the hum of the collaboration starts and collective learning follows. And then collaborative action!

Your doubts and fears are echoed by many first-timers to the collective learning process. The collective learning cycle looks simple and much like other collaborative processes at the start. This of course allows the participants to feel relaxed. However, as you found, every step is quite different from the normal run of change management process. Beginning with accepting a diversity of values, then going on to accept a diversity of facts, sets a different social and intellectual environment that allows for, not just incremental, but transformational change.

The rejection by well-qualified and skilled facilitators of the old school is also a familiar response. Trained to supply firm frameworks and set clear objectives, it can seem heresy to start where the clients, not the facilitator, are at, and let them set the goals and develop unexpected outcomes.

And finally, it is not unusual to find the most outspoken critic becomes the strongest advocate. I explain this as the criticism fighting the transformative insight – and when it comes you find you have a disciple!

I hope you will keep us informed about how the changes go forward from here.

Warm regards

Valerie Brown and Judy Lambert

Figure 19.1 Raw data, as collected at the Montessori workshop

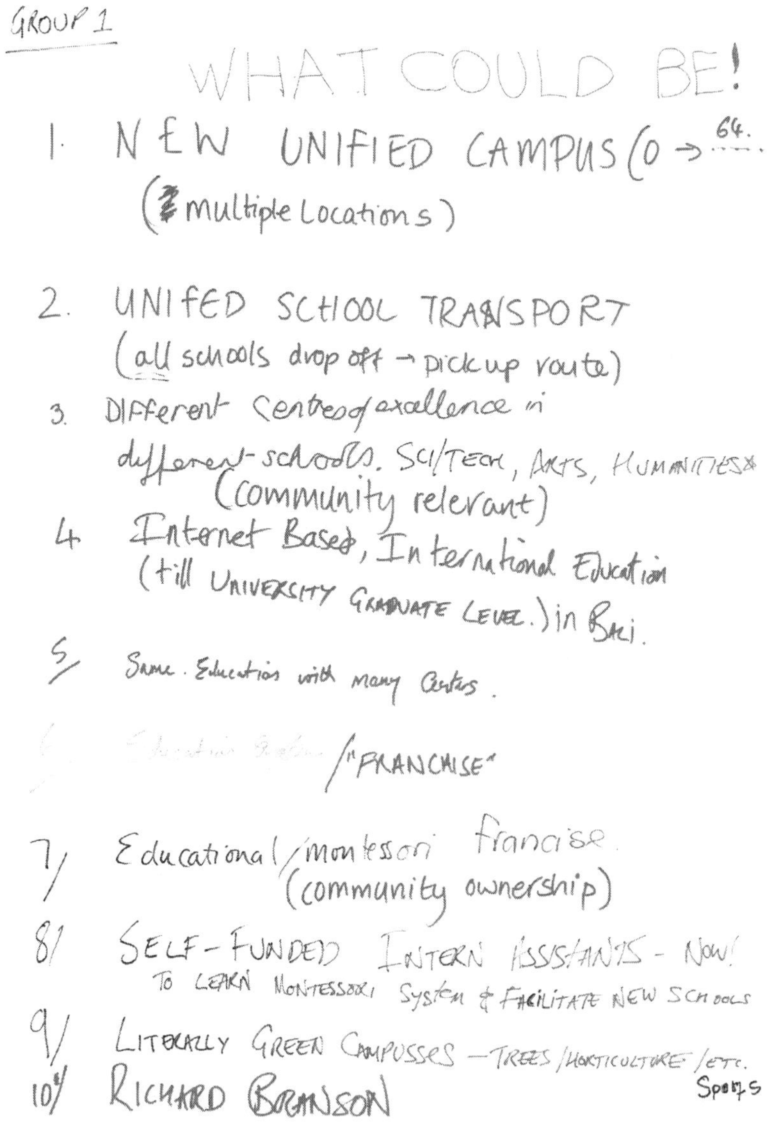

GROUP 1

WHAT COULD BE!

1. NEW UNIFIED CAMPUS (0 → 64)
   (& multiple Locations)

2. UNIFED SCHOOL TRANSPORT
   (all schools drop off → pick up route)

3. Different Centres of excellence in different schools. SCI/TECH, ARTS, HUMANITIES
   (community relevant)

4. Internet Based, International Education
   (till UNIVERSITY GRADUATE LEVEL.) in Bali.

5. Same Education with many Centers.

6. Education ?.../ "FRANCHISE"

7. Educational/montessori francise.
   (community ownership)

8. SELF-FUNDED INTERN ASSISTANTS - NOW!
   To LEARN MONTESSORI System & FACILITATE NEW SCHOOLS

9. LITERALLY GREEN CAMPUSSES — TREES/HORTICULTURE/etc.
   Sports

10. RICHARD BRANSON

Working from the Guidebook

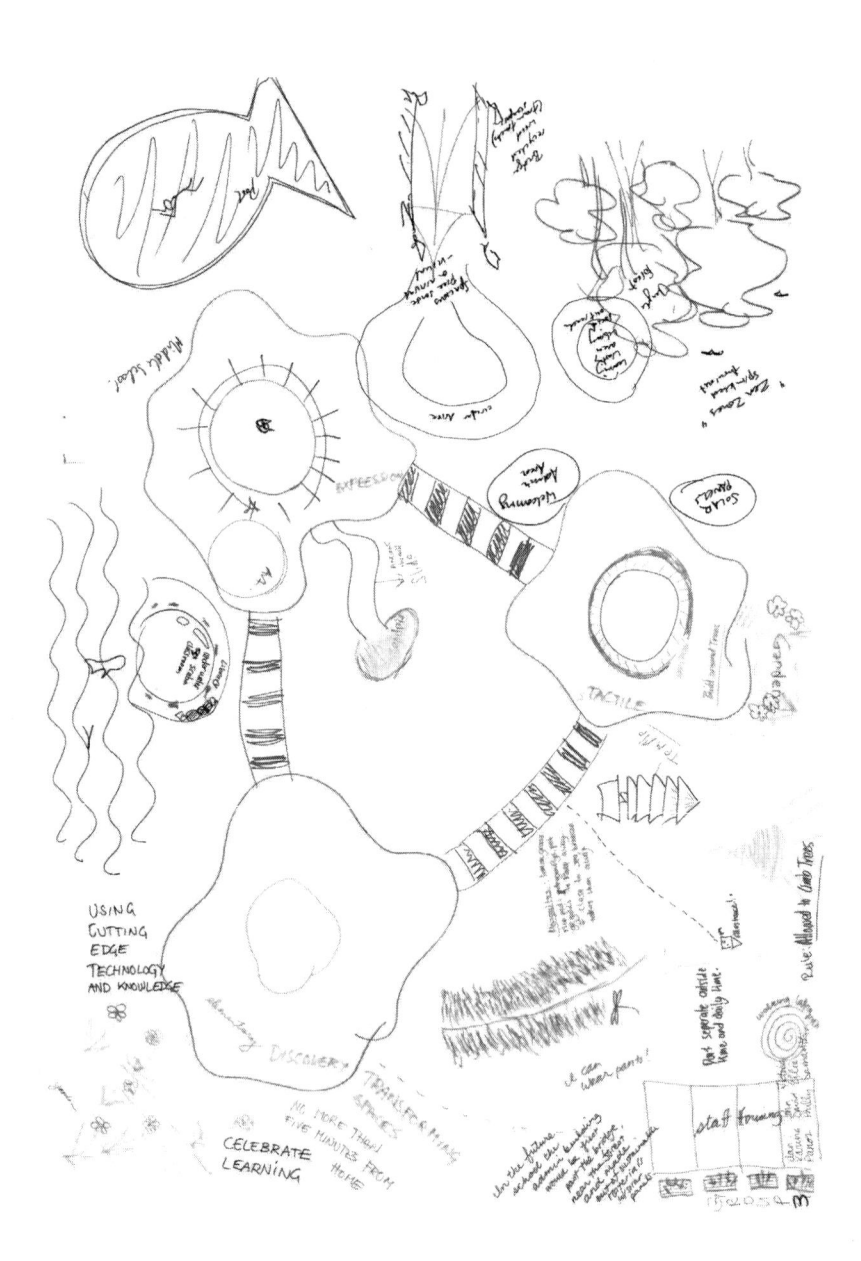

200 Collective Learning for Transformational Change

Name	Commitment (action)	By When? (Date)	Resources
Eve Kelynn Two more Parness	Spread the word of Montessori	immediately	give, and people to self smile ☺
John i Sullivan	2 classes per term to facilitate	13/1	IAPC, NAE, ION + Joyous Students!
John, Philly + David + Charles	Meeting to discuss structure i effective management + support systems to reality shape and to defined group but policy + lists to opposition	8/12/11	Busy mother
Ian Thompson	Facilitating the development of a master plan, training goals etc. as needed	open ended	C.F MILLER ARCHITECTS
Karine	Continue my mission as a Montessori teacher, follow my talents, inspiration	ongoing	me, children, my assistant fellow teachers, staff
Samantha	· Inviting speakers, Montessori educators to talk to existing community · Use a manifestation of "collective learning" "a different way board of school the children deserve" drawing a part of	encouraging Nov 12/13	teachers Montessori talk students at LE2
Victoria	Share the case study of TGM's new low-energy school office with Montessori staff to share learning	16/12	n/a

Samantha Ellen
Ian Karine
John Philip
Victoria Charles

2. CASE STUDIES

What could be? ③
IDEAS | PROJECTS | POSSIBILITIES

- Generous plot of land.
- 30 mins from Seminyak Area.
- Toddler - Middle school
- Capacity for further growth
- Local community engaged respected
- Research + incorporate environmentally friendly materials
- (Climate controlled rooms - fluid spaces)
- Natural design with good sense of flow.
- School bus
- Productive gardens
- & Self autonomy - energy
  - food
- Inter-generational shared learning
- Parent involvement drawing on parent knowledge = knowledge sharing
- INSIDE/OUTSIDE FLOW DESIGN
- CLASSROOMS (a) aesthetic (b) functional (c) educational
- ZEN ZONES
- SENSORIAL MATERIALS

- includes "beautiful, rejuvinating" staff housing/space
- Sense of arrival
- Sports space/teams/ Pool Competition, BSSA sports
- Concerts/theatre multi-purpose hall
- SUSTAINABILITY IN EVERYTHING

*Life cycle considerations

*Home Economics
 - cooking
 - sewing
 - wood work

• Classroom pet that kids take care of

Samantha
Ian   John
Lavinia Philly
Panos Victoria
David Ellie

• WATER/WASTE MANAGEMENT

*Designed for Children

Enablers (+) | Disablers (-)

In Bali
- incredible Resources
- nature / spirit

Finance (access to)

Passion

Location - great / possible location

Marketing/Educating about Montessori

Community in place

Montessori Staff

ONE SCHOOL BOARD

Parents believe that they are investing in their children's education

International curricula can be used here

Progress / more acceptable / understanding that holistic education works... well! series

Appealing Place to live

Local culture (conflict)

Finance (access to, no X)

Societal expectations (difference one makes between two worlds)

Poor infrastructure

"Extra" classes (P.E., art, etc.)

The school boards

Steve / Martin
Simon / Donna
Phillip / Josh
Hans / Kim
Karl

Name	Commitment (action)	By When (date)	Resources
1. DEWI	Be an efficient & effective secretary of Montessori Elementary school	today	my office, telephone
2. Dilma	Investigate Tamman Petmua	3 months	Donw/can
3. Gavin	To connect the school to Ubulistic Business Network and Tamman Petmua	tomorrow	Email
4. Dodi	Will nominate yoga session so children will get benefit & have fun in doing yoga learn	Monday	Books
5. Wilma	Hold education/discussion evenings. To introduce changing educational paradigms. To motivate parents.	3 months	petra projector — video — powerpoint information.
6. Nathalia	Investigate Tamman Petmua with Wilma	3 months	Time
7. Nathalia / Kadek	Look for Land!	Within 3–6 months	Time, money, resources
8. Victor	Continue to teach with passion	more almost to grow for 7 more years	me + my class
9. Chiemi	Support as mush I can! I have a book about Montessori in Japanese. If copy and give to school.	Next week	me.

Case Study 16
Developing a strategy for regional action on health

Story: US State Department hosts a planning workshop for transformational change in regional Ethiopia

Transformation: From starving children to a well-fed and healthy community

Collective: Greg Bruce, Humberto Swann, Aleksandra Moiseeva, Ahmed Abu Al-Halaweh, Karim Khelifi, Vitaliy Shuptar, Rabab Rmaini, Nagaddya Daisy, Ian Fitzgibbon, Rabii Larhrissi, Imane Jroundi, Timur Mykhailovskyi, Patricia Ndele, and US Workshop Host – James Bunch.

Learning: Single turn of collective learning cycle

Leadership: Greg Bruce and James Bunch (US table host) Professional Fellows – US State Department Congress 19–21 October 2011

Context: A group of UN professionals, at least one of whom was familiar with Collective Learning for Transformational Change, sought to develop a strategy to reduce the prevalence of acute malnutrition in Oromiya Region of Ethiopia. This group decided to use Brown's collective learning cycle to develop a strategy and actions to address this issue. The material that follows was provided by the project leaders. It is a record of the project's progress, as recorded by participants.

> *Focus Question:*
> *How to reduce the prevalence of acute malnutrition in Oromiya Region of Ethiopia?*

Developing a strategy for action using collective vision and actions

19–20 October 2011

Professional Fellows of State Department (Group 9)

Workshop participants: Greg Bruce, Humberto Swann, Aleksandra Moiseeva, Ahmed Abu Al-Halaweh, Karim Khelifi, Vitaliy Shuptar, Rabab Rmaini, Nagaddya Daisy, Ian Fitzgibbon, Rabii Larhrissi, Imane Jroundi, Timur Mykhailovskyi, and Patricia Ndele (Professional Fellows 2011 – Group 9)

Disclaimer: The following information represents the individual truth and experiences of each participant.

The objective was to generate collective and individual thinking and communication.

The content as recorded by group participants was compiled and the workshop facilitated by Greg Bruce and James Bunch (US table host) Professional Fellows – US State Department Congress 19–21 October 2011.

The ideas from each group of what should be, what is, and what could be

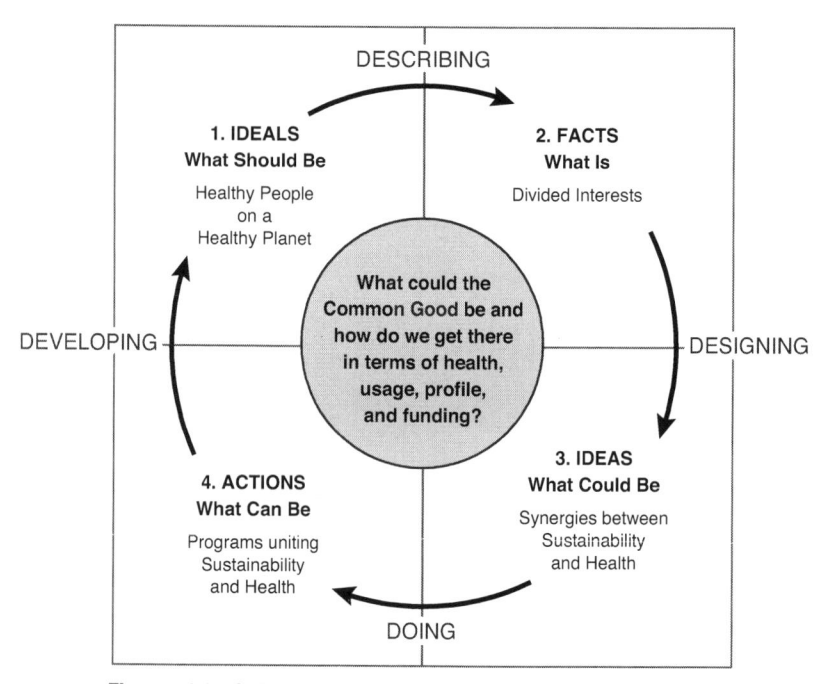

```
                        DESCRIBING

        1. IDEALS                    2. FACTS
        What Should Be               What Is

        Healthy People               Divided Interests
           on a
        Healthy Planet
                    ┌─────────────────┐
                    │ What could the  │
                    │ Common Good be and│
DEVELOPING          │ how do we get there│         DESIGNING
                    │ in terms of health,│
                    │ usage, profile,  │
                    │ and funding?    │
                    └─────────────────┘

        4. ACTIONS                   3. IDEAS
        What Can Be                  What Could Be

        Programs uniting             Synergies between
        Sustainability               Sustainability
        and Health                   and Health

                          DOING
```

Figure 19.2 Collective Social Learning process for workshop.

What Should Be?
A vision for 5 years from now (how would it be?)

- Educated mothers
    - breast feeding
    - hygiene
    - detect early signs of malnutrition

- Sanitation and water supply generalized in region

- Health community workers engaged and competent
    - education
    - prevention
    - advisors

- Local production of RUTF

- Reducing underweight children from 34% to no%

- Diet health education in schools

- Improved accessibility to facilities (road; transportation)

- Creating opportunities of work and well developed agricultural infrastructure with educated specialists

- Reduced corruption level in the area of distribution of humanitarian aid – food, medical services

- Strong and sustainable social help projects (both funded by State and international organizations

- Civil society involved in social problems

- Community based tourism network is working for 2 years already, providing additional income directly to the rural communities

- Reduced the unemployment rate from 26% to 0%

- Have a solid higher education system

- Steady households to enable people to provide them with food – creating jobs and making the region attractive for investing

- Legal support of agriculture activity, taxation, involving people, exporting technologies and medicines

- System of additional support and nutrition of children in schools and kindergartens

- Support for better family planning

- Implementing evidence based treatment strategies

- Conducting training and seminars for interesting group of people

- Systematic collaboration between government and NGO

- Good partnerships with Corporate bodies for food resources and health

- Seeds and water supply to local farmers

- Strong local leadership in community

- Providing community training
    - in agriculture
    - health

- Agriculture production sustainable model (farmer – farmer)

- Diverse employment across local economy for man and women

- Community-based programs / different approach

- Community awareness

- Empowerment of women / literacy

- Train village health team

- Think of the problem from a wider scope, vision and consider
    - Legislation
    - Political
    - Economic
    - Social
    - Cultural
    - Religion
    - ....

- Supplying nutritional seeds to community

What Is?

"How to reduce the prevalence of severe acute malnutrition in Oromiya Region of Ethiopia"
Developing a strategy for action using collective vision and actions

Sheet one
- Health, education – low cost program
- Donors/funding
- Community involvement

Sheet two
- Micro-finance
- Leadership for youth and women
- Higher unemployment rate
- Rural way of life is interesting to foreign tourists
- Political will
- Development of partnerships and with corporate bodies to supply resources i.e. milk powder

Sheet three
- Devoted/dedicated people in community
- Evidence based experience from other countries
- Responsible attitude of people to their life
- Responsible government
- A lot of hidden resources available in the community

Sheet four
- Share experiences
- Share responsibilities
- Involve many partners
- Water resources preservation (potable-waste)
- People support
- Understanding of reforms

Sheet one
- Corrupt political systems
- Lack of schools, trained health educators
- Implementers with different interests
- Individual needs and survival

Sheet two
- Poor transport network (Roads and Ports)
- Funding/Corruption
- System of education and vocational training are not aligned with the market needs
- Language barriers and lack of possibilities for improvement
- Corruption of officials
- Lack of corporate social responsibility

Sheet three
- Support not based on strategic planning
- Lack of follow up and supervision
- Lack of awareness of communities and people
- Division / Cloudy role of health clinics in community "illiteracy"

Sheet four
- Lack of budget funds ($1m)
- No political will
- Afraid to lose independence
- Lack of consciousness of people urban – rural areas
- Political changes
- Not understanding

What Could Be?

- Develop global philanthropic project

- Encouraging identification of leaders for, education, advocacy, negotiation with industries and governments

- Clearly identify the streamline for education in agriculture and health the streamline from up to bottom and bottom up

- Develop partnerships with international milk producing companies to supply free milk powder to use in RUTF

- Involve teachers and health professionals in educating mothers and children about nutrition

- Community based tourism development project is implemented by an international and national NGO focusing on education and local communities capacities building and promotion of the country abroad

- Milk powder distributed at the local level (not national) by locally elected representatives

- Agriculture sustainable programs (technical assistance)

- Community based project target, awareness women empowerment, health and lifestyle

- Schools, where children are taught initiative and responsible / general education

- Comprehensive health care services

- Create educational centres in rural areas in order to provide the basic education of students who drop out from school to work and the families

- State Department use all of us to help articulate this plan with assistance of James (from Oklahoma University)

What Can Be?

Name	Commitment/ Action	By When/ date of action	How	Resources
Greg	Email James	Next week	Email	James
James	Contact State Dept.	Next week	Email	Curtis
Ahmed	Share the women model with everybody	Dec 2011	Email to group	Email list
Karim	Identify local community leaders	2 months	Email	Email list
Vitaliy	Contact NABU	Next week	Email to group	Email list
Rabab	Prepare budget proposal	Nov 2011	Email to group	Email list
Humberto	Technical assistance to agriculture	Nov 2011	Email to group	Email list
Alex	Share the idea	Today	In conversation	
Daisy	Model woman	Dec 2011	Training	Email
Ian	Contact friend in milk company about the project, issues and possibly donate milk powder for RUTF	Nov 2011	Email	Communication
Rabii	Health education model	Nov 2011	Email	Group email list
Imane	Health and educational support doc, video	Nov 2011	Email	Group email list
Timur	Ask for help	Immediately	Mass media	Communication
Patricia	Sensitize community	Dec 2011	Conduct open air MCH clinic	
Karin (supporter)	Point of Contact, with Dept of State, US Embassy in Ethiopia, Australia, State ECA, Ethiopia, Alumni and list of Public Private Partnerships	Long term	Via email and phone	

Working from the Guidebook   211

What Can Be?

Project Idea PLAN	What	When (date of action)	Resources
Integrate links, synergies and agree the plan	From the project and ideas (What Could Be) develop agreement on which projects could be possible and how they connect. Integrate linkages and create synergies between divergent sectors and ideas.	Thursday 20 Oct 11	Team and work room
Write up Plan	Produce one page summary based on agreed plan	Thursday 20 Oct 11	Team and work room
Present the plan	Present plan to all State Department Professional Fellows	Friday 21 Oct 11	Professional Fellows congress
Individual plans	Individual agreements and actions to be undertaken by participants	Various	Individual and agreed

Following on

The international working group that participated in the workshop planning and delivery agreed to continue to act as a support group. The in-country group agreed to evaluate the planned projects and lobby local governments to implement the plans.

Project Contacts

Name	Role	Country
Greg Bruce	Executive Manager Local Government – Sustainability	Australia
James Bunch	University Representative	USA (host)
Ahmed Abu Al-Halaweh	Director, Department of community programs and the diabetes care center	Palestinian Territories
Karim Khelifi	Management / lecturer / marketing	Algeria
Vitaliy Shuptar	President Historico-geographical society	Kazakhstan
Rabab Rmaini	English teacher (program coordinator)	Morocco
Humberto Swann	General Manager Environmental Engineering	Colombia
Aleksandra Moiseeva (Alexandra Moiseyeva)	Lawyer at the Public law department	Russia
Nagaddya Daisy	Community (farmers development association)	Uganda
Ian Fitzgibbon	Manager Local Government – Sustainability	Australia
Rabii Larhrissi	Doctor – primary care health center professional	Morocco
Imane Jroundi	Epidemiologist / teacher / researcher	Morocco
Timur Mykhailovskyi	District council member, political consultant	Ukraine
Patricia Ndele	University Lecturer	Zambia
Karin	US State Department (Europe/South and Central Asia Division)	USA (host)

Working from the Guidebook

20 Summing up

Summary: This chapter takes a look back at what has been learnt from use of the collective learning spiral in a diversity of case studies.

Source: Art of Moving (2006)

The pressure for world-wide collaborative action towards a sustainable future has been building throughout the 1990s and early 2000s. In 1992 the first international Environment and Development Summit was held in Rio de Janeiro, as a collective consultation with scientists, heads of state, local government and community representatives. In 1999 the joint American Academies of Science published *Our Common Journey* with a strong plea for a sustainability science which integrated science with other ways of knowing. Fast forward and in 2011 the new governor of the World Bank announced a policy of collaboration with all the Bank's clients. In between were numerous summits and peak councils, increasingly with a wide range of representatives instead of the usual suspects: heads of multi-national institutions. A United Nations investigation on the future of big dams produced a prototype of collective learning as the basis for multi-national planning.

Then came Copenhagen and the heads of state worldwide tried to find a way forward to counter climate change. The meeting split between developed and developing countries with different agendas. All the hopes of a constructive collective were dashed. It has since been since said that at least the real agendas were on the table. However, that is poor comfort when so many skills and so much energy had gone into a global social learning exercise.

The case studies of transformational change in this book tell a different story. All grounded in the local, from large regions to small groups, collective learning has successfully supported large (a whole city) and small (a school board) transformational changes. The changes described here are mostly concerned with countering environmental degradation and moving towards a sustainable future. Each documents an intentional transformation which at the time of writing was continuing along the collective learning spiral.

SALLY AND RICHARD 9

Are we there yet? Reflecting on a shared journey

Five years on from their initial collective learning project for regional change, Sally and Richard met up again. Both had moved on to another phase in their lives, but both were keen to reflect on the journey since that early initiative.

Richard observed that their program had spread over three continents and that many of the projects arising out of their initial work were still going, many now run from within their own communities without the need for external facilitators. For some, the collective learning cycle has become so much a basic foundation of everything they do, that they no longer remember how they started.

The diversity of people and situations in which the cycle is being applied seems almost infinite, reflecting the ways in which people and their settings differ. Both Sally and Richard reflected on the extent to which that richness and diversity had provide so many learning opportunities for them.

Although David Kolb's adult learning cycle provides a basis that worked in every case, the challenges faced in building on that cycle and fitting the shared changes back into the settings from which they had been developed were often significant.

For Richard, the spark of excitement generated when different people come together to create change through a collective process was a source of continual enjoyment.

Sally acknowledged that both she and Richard had become better at helping with that transmission as they gained experience with the process.

Sally and Richard felt pleased that they had reached their goal of embedding the collective learning spiral in community change processes. However, they also reminded each other that the journey is not over yet. Collective learning is an idea whose time has come.

The case studies in Part 2 are only 16 of more than 300 uses of the collective learning spiral, and we are still counting. There are more because many projects continue on in ripples from the original learning cycle and vanish into the distance. It seems time to do a quick review of the learning from the scope and diversity of the projects and programs we do know about.

The collective learning spiral with its multiple knowledges has been introduced successfully into at least 12 countries (Australia, New Zealand, Malaysia, Nepal, Fiji, Canada, Holland, Britain, Ireland, Ethiopia, Kenya, Namibia). When used in a non-Western culture it seems that the learning cycle is not cultural colonialism in itself, but rather a framework into which any culture can insert their decision-making system. The content of the cycle is always re-designed to fit every application. The framework remains the same.

This has allowed people from traditional cultures to frame their own knowledge in their own terms. Users of the framework in developing countries have used the cycle of Western knowledge to identify how the power of Western knowledge from education, research grants, and specialist advice from all sources, can have the same effect as colonization.

Each use of the spiral has demonstrated the power of difference. Not only did the use of the spiral open the way for desired transformations. It moved the idea of transformation as a struggle to the idea of transformation as a celebration. Almost all the case studies identify a sense of excitement, exhilaration and achievement.

Step 1. Setting the scene

The learning cycle was originally designed as a closed cycle of individual learning. The present study moved its application to a collective decision-making spiral, a move shaped by experience in the field. All human learning is cumulative, that is, each learning step builds on the learning from the one before. This was equally true for the teams who mounted the projects and for the authors of this book. In almost all cases the cycles build into a spiral.

In-the-field learning included the crucial need for a focus question for each learning cycle. The question needed to be workshopped with the potential participants, to make sure it was relevant to everyone involved. This did not mean that the answers to the question or the associated facts, or even the meaning of the question, was agreed to. It did establish the common ground of a mutual interest that could start a learning cycle.

> *Focus question:*
> *What have we learnt from the range and focus*
> *of the learning cycles of the collective learning spiral?*

Step 2. Stage 1 of the cycle

Ideals: Success in placing ideals as the initial step on entering the learning cycle was a key to the success of the whole cycle. Acknowledgement of and respect for each other's ideals led directly to mutual trust and to the recognition that the process was a process of change. Several projects began by people objecting strongly to beginning with values rather than facts. 'We always begin with the facts' was a common cry.

After a cycle is completed, participants recognize that this beginning allowed for change. Starting with 'the facts' would fix the focus question in the present issues, when the object of the exercise was transformational change.

Many groups were so unused to having each of their ideals given equal value that they regularly asked 'Are we going to prioritize these?' 'How are we going to reach a consensus?'. They were quite surprised when the response was 'This is a collective exercise. Everyone's ideals will be equally valued all the way through'.

Ideals could take the form of value statements, aims, purposes, visions and hopes. It was often necessary to explain that we were not asking for objectives or wish lists. Nor are we seeking the predetermined outcomes beloved of program terms of references. By definition, the outcomes of a transformational change cannot be known at the beginning.

Many exercises were used to ensure that this stage remained focused on ideals and that everybody's ideals were heard and respected. The most effective tool was the visioning exercise, (see Visioning in Part 3) described in Case Study 1. A wide range of other tools for values clarification can be found in Part 3.

Step 3. Stage 2 of the cycle

Facts: In practice this was the learning stage that led to the most debates and even to real conflict. In Stage 1 people seemed willing to respect each other's ideals, no matter how different. In Stage 2 the participants were again asked to contribute individually, this time naming the supporting and impeding factors to achieving their own ideals. When this stage was managed as information-sharing, then differences as to 'the 'facts' could be resolved. When 'facts' became confused with right and wrong, good and bad, this then challenged the values expressed in Stage 1.

The collective learning framework accepts that a factor of importance to any one of the participants was an important factor in the decision process. The same factor could be both supporting and impeding. Different factors were important to different people. It was accepted that 'truth' was a matter of 'whose truth?' and tested by different criteria for each of the multiple knowledges (see Chapter 2).

It was less than desirable, but sometimes necessary to describe the factors affecting the change process through pre-existing documents. These descriptions are often from only one of the multiple knowledge perspectives, so be careful to ensure that multiple information sources are included.

The most successful tool for this stage was the Kurt Lewin Field Force analysis described in Chapter 6. Other suitable tools from Part 3 include Dialogue, Knowledge brokering and Yarning, which help build shared understanding of sometimes very different 'facts'. Forecasting provides a fact-based approach to the future, and it is at the 'What is?' stage of the cycle that participants are most likely to need some skilled conflict resolution.

'Counting the things that can be counted, and accounting for the things that can't be counted.'

Step 4. Stage 3 of the cycle

Ideas: While the first two learning stages are made up of individual positions, the next two involve the full collective. Stages 1 and 2 have allowed everyone to understand each other and to scope the full range of the possible transformation. In Stage 3 the team-building proper begins. The linking exercise between facts and imaginative ideas is particularly important here. These exercises have included a physical activity (handball), a living sculpture, music (in one case Indian drumming), constructing and reading haikus, a reflective circle, a convivial meal, sharing personal experiences ('where were you when... ?' or 'share something surprising about yourself'), a bush walk. Such activities help the participants make a shift from self-oriented to shared thinking.

In Western cultures the use of the imagination and intuition (holistic thinking) is often downgraded and not included in explicit discussions of how decisions are made. It seems that it is not culturally acceptable to say 'I used my imagination' and 'I trusted my intuition'. Yet for any transformational change this implicit knowledge we all carry is as crucial as any other form of knowledge. It allows ideas to emerge that have been suppressed by the dominant cultures of specialized and strategic knowledge.

Step 5. Stage 4 of the cycle

Action: This Guidebook is built on the commitment and hard work of collective learning practitioners working with communities, professions and government agencies committed to transformational change. The authors hope that the readers will carry the ideas forward.

Step 6. Following on

Throughout the use of the collective learning cycle we have assumed that for transformational change all the knowledge cultures needed for collective decision-making are made explicit. The release of ideas that do incorporate all the knowledges is crucial for the ideas to be both effective and supported. Mixing knowledge cultures is the key to generating ideas 'outside the square'.

The conditions that release imagination and intuition need to be carefully considered for each separate case. Indeed, this stage calls on the imagination and intuition of the hosts. Fresh 'blue sky' ideas are more likely to emerge where there is both trust and challenge among a group. This requires careful scene setting. Useful tools in Part 3 include the Xing a minefield exercise, which helps participants understand how others perceive a challenging situation and how they might react to it. Imagining and strategic questioning, both of which encourage participants to consider all alternatives are also useful. A capacity for risk-taking is necessary if the group is to move beyond 'What is?' at this stage.

Figure 20.1 All together now!

PART 3.
RESOURCES

A–Z of Collective Learning

Introduction: Who? Why? What? When?

Parts 1 and 2 of this Guidebook explored the role of the collective learning spiral in transformational change. The problems to be tackled in transformational change are complex, challenging and require social change. Special tools may be needed to assist with the change and at the same time make it fun.

Part 1 describes the six steps in an effective collective learning process: setting the scene, developing ideals, describing facts, designing ideas, doing in practice and following on, while enjoying the collaboration.

Part 2 gives examples of taking those steps, presented as eight different types of celebration presented in 16 case studies. These examples confirm that any one of the steps may need you to select special tools that enable the diverse interests to work together in a party spirit.

Part 3 helps you to find the right tool for the right job at the right time. The diverse set of tools is presented as an A–Z of collective learning tools. For every step of the collective learning spiral, the choice of a tool should take account of the answers to the questions: Who? Why? What? When? Variations on each tool can readily be found using the numerous search engines. At least one resource is included for each tool – where references are not included in full in the tool section, they can be found in the bibliography. The versions of the tools given here are discussed in relation to collective learning and transformational change.

Who can use the tools? Answer: all those involved in a collective learning process. A transformational change requires whole-of-community support if the change is to be established over the long term. Collective learning generates that support through the collective contributions of key individuals, affected communities, relevant specialists, influential organizations and creative thinkers over four learning stages (see Knowledge brokering and Multiple knowledges).

Ideally, each collective learning cycle should involve a full set of these interests. Since recruiting all the interests is not always possible, the collective learning process should involve as many different views as it can. To repeat Bohm's edict: 'We learn from difference not from more of the same.'

Why use tools? The next question is why use a tool at all? Why not ask the learning group the direct questions: What are your ideals? What are the facts? What are your ideas? What action should be taken? Answer: the current division of interests is so complete that some remedial action is needed for them to work collectively. Participants have different learning and communication styles and these need to be made transparent so they understand each other. The right tool also deepens the level of response and increases the enjoyment in the process. The wrong tool can change the atmosphere and lead to mistrust.

What tools for which job? Each collective learning spiral is grounded in a particular place with particular interests. It is important to select tools that best match the needs of each case. If a tool is pre-selected without due consideration of the context, the program risks the fate of 'when your only tool is a hammer, all your problems will look like nails'. The tools in Part 3 cover a broad selection of those the authors have used in the field, with a commentary on their suitability for collective learning.

TRUST built on - Communication - Honesty and Openness - Cooperation - Experience	**MUTUAL RESPECT** for - Skills - Knowledge - Needs and Expectations - Values
EQUALITY / POWER-SHARING of - Opportunity - Ideas - Effort - Decision-making	**COOPERATION** leading to - Economies of Scale - Better Decisions - Shared Workload - Celebration of Diversity

Centre: Collective Learning & Action

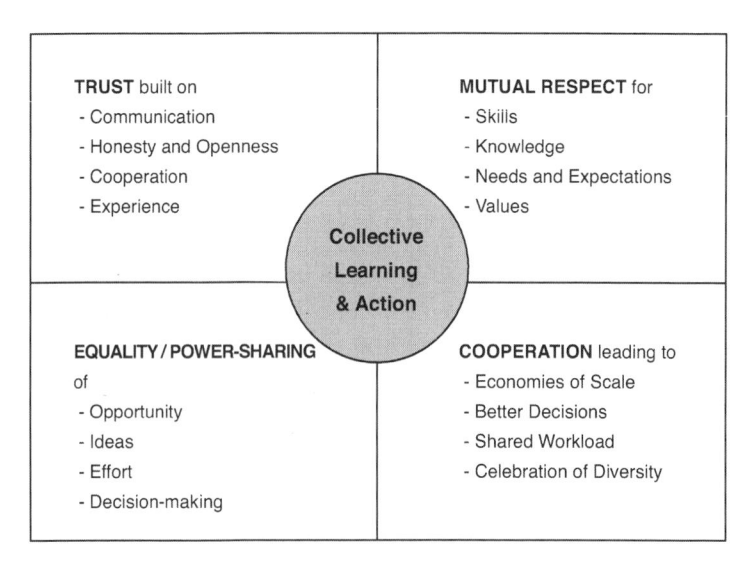

Figure A–Z.1 Building collective learning and action
Source: Elix and Lambert (2007)

When can we use the tools? The tools are cross-referenced to the six steps of collective learning. Many of the tools are simple and direct and can be employed by anyone in the collective learning team. Some need experience in facilitation and negotiation. Others are more complex or more sensitive and need further information and possibly some practice. Common sense decides which is which. All of the tools can be explored further through internet search engines such as Google.

A collective learning approach expects that the leaders as well as the participants take part in transformational change. One of the authors has found the framework in Figure A–Z.1 useful when thinking about the tools that might assist in achieving such change. You may find it of value as a discussion tool among your collective learning team.

Adaptive management

Why? In the 21st century, organizational management has moved away from top-down fixed management styles with pre-set objectives, performance and outcomes. In times of uncertainty and constant change, most managers now recognize that action on any complex issue needs to be responsive to learning from expected and unexpected events, to include the organization's clients and community, and to adapt their management accordingly.

What? Adaptive management has evolved along several lines, depending on the field of management and the personality of the managers. In natural resource management, adaptive management is defined as an approach that involves learning from management actions, and using that learning to improve the next stage of management. The symbol is the never-ending Mobius strip (Holling, 1978).

The Resilience Alliance approaches socio-environmental change as a system that can respond to shocks and return to its original state, and so act as a resilient system. More extreme shocks that change the system are said to cause a regime change (Walker and Salt, 2006). The considerable amount of work with resilient systems is an adaptive management cycle.

For the New South Wales Natural Resources Commission and others, adaptive management is similar to the collective learning cycle in its strategic planning form, although it leaves out the initial step of clarifying ideals, and is represented by a closed cycle rather than an open spiral.

Figure A.1 Standard adaptive management cycle
Source: NSW Natural Resources Commission.
www.nrc.nsw.gov.au [accessed 8.7.12]

Forward-looking organizations increasingly couple adaptive management (learning while doing) and co-management (involving all interests in decision-making) (Armitage et al., 2007). Other features include multi-level and multi-focus governance. This approach has lots of lessons for putting collective learning into practice.

When? Adaptive management ideas fit into the collective learning cycle at Stages 3 and 4, ideas and action. Ideas from the literature on adaptive management can open up new avenues for managing change. Techniques of adaptive management can form part of the toolkit for collaborative action

Alliancing

Why? Alliancing (formal alliances between participating organizations) has become a way to design major engineering projects faced with high levels of uncertainty. Alliancing involves the partners in a project developing shared goals, collaborative practices and negotiations that seek gains for all involved, while at the same time ensuring shared risk. This is in marked contrast to the traditional competitive contracting. The idea of alliancing can be extended from business to governments and organizations (see Case Study 6).

What? The stages in alliance formation can typically be described as:

- Strategy development
- Partner assessment
- Contract negotiation
- Alliance operation
- Alliance termination

One alliance award-winning project is RIAMP (Reliability Improvement and Modernization Program) in Sydney, Australia. The citation for the alliance award described RIAMP as successfully carrying out $A75M of upgrade work on an underground sewage treatment plant in operation 24 hours a day, seven days a week. This professional, environmental and financial success rested on an alliance based on open dialogue among project managers, continual engagement with all the interest groups, and an emphasis on transparency and teamwork (Russell Greene, Quality Project Manager, RIAMP, pers. com.).

When? To date, alliances have primarily been used in the construction sector. Their relevance to complex issues in other fields is reflected in change in contracting procedures in social service organizations. A short list of bidders are given each other's proposals and asked to put in an amalgamated bid.

Alliances are developed in ways that enable each of the participants to remain independent and retain their own identity, while working as one team striving for shared outcomes. They require skilled leadership and an organizational willingness to change. Given this climate, alliances can foster self-reliance and a culture of 'innovation and excitement'.

'The Alliancing Association is an independent, not-for-profit cross-industry initiative created by practitioners for practitioners to develop their excellence in business collaboration. AAA is focused on providing a forum for those individuals and organizations who are or want to have involvement in alliances so they can leverage a pool of knowledge to assist them to:

- Strengthen their alliancing and collaborative contracting competencies
- Voice their ideas, concerns on the challenges and the practices of Alliancing
- Assist in the evolution of alliancing models in consideration with changing business environment'

See: Alliancing Association of Australasia.
http://www.alliancingassociation.org/index. [accessed 18.1.12]

Balancing the players

Why? Traditionally Western culture has placed greater value on specialized (technical) and organizational knowledge and their languages than on individual or community knowledge. As well as this imbalance, individual personalities enable some to express themselves more readily than others. For complex problems all voices need to be heard if there is to be all-of-community support for transformational change. This means remedial action is needed to ensure balanced, effective decision-making.

What? The five different knowledge cultures (see Knowledge brokering and Multiple knowledges) provide a basis for deciding who needs to be invited for a balanced collective learning process. However, individuals can be simplistic, communities fragmented, specialists one-eyed, organizations self-serving and holistic thinkers unrealistic. The collective learning cycle is designed to balance these biases without reducing their diversity.

Several authors offer schemes for helping to balance the personalities as well as their knowledge bases. Kolb gives us learning styles (L). Keirsey (1998) identifies people according to their long-term personality traits, so they can be allowed for:

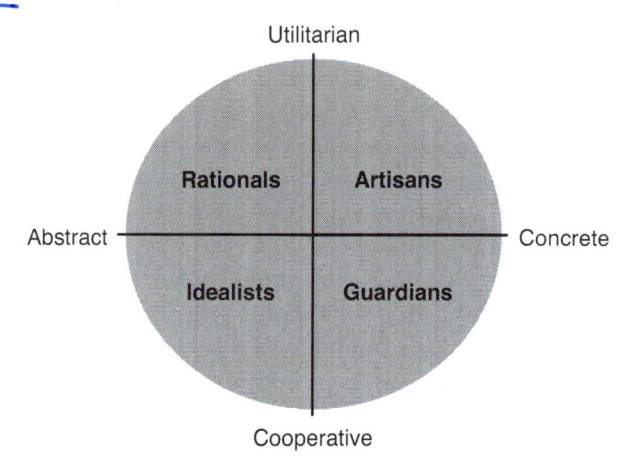

Figure B.1 Keirsey's four major temperaments

The Myers-Briggs Type Indicator groups people by how they approach life. Each person can identify their own approach through choosing from four pairs of alternatives and share this with the others in the group (e.g. ISTP or ENTP):

E = Extroverted	I = Introverted
S = Sensory	N = Intuitive
T = Thinking	F = Feeling
J = Judging	P = Perceiving

When? The relationship between participants in a collective decision-making process changes over time. As shared understanding and trust are built among the participants, it is important to ensure that balance is maintained across individuals, knowledge types and decision-making power throughout all stages of the collective learning cycle.

Also see: HumanMetrics.
http://www.humanmetrics.com/cgi-win/jtypes2.asp [accessed 18.1.12]

Collaboration

Why? Complex problems have complex answers that need to involve many interests if the solutions are to be successful. A respect for collaboration is central to collective learning. Without a spirit of collaboration the contributing interests will not even listen to one another, much less enter into dialogue (see below). A transformational change initiative does not necessarily start with a spirit of collaboration; this will need to be developed during the process. Dictionary definitions of collaboration include the act of working with others on a joint project, something created by working jointly with others, and the act of cooperating with the enemy. These definitions emphasize that collaboration can be a process or an outcome and can be positive or negative, depending on how it is applied in practice.

What? Collaboration in practice is often surrounded by the halo of 'all collaboration is a good thing'. In a competitive society this is often not the case; simply putting people together can increase the tensions between them. Collective learning depends on establishing constructive collaboration, which takes commitment and effort on the part of all concerned. Constructive collaboration may involve conflict resolution and/or negotiation (see below) where there are fixed positions that need to listen to each other.

Activities that build collaboration are an integral part of each of the six steps of the collective learning cycle. In this Guidebook there are 16 examples of how different initiators of change developed a collaborative environment. The collaboration in every case involves the key interests of individuals, community, specialists, organizations and creative thinkers (see Chapter 4 and Multiple knowledges). In addition, the considerable body of work on principles of community collaboration is an invaluable resource. Authors to access for detailed descriptions of community collaboration as a change management practice include Wendy Sarkissian (2010) who offers a creative approach for taking a community into a new space and Julia Wondollek (2000) who applies community development principles to innovation in natural resource management.

The review of learning styles in Chapter 10 is another resource for establishing a collaborative team. Within a collective learning spiral, teamwork expands as time goes on. Participants with contrasting views can develop strong partnerships; existing colleagues can develop their shared ideas even further. After a couple of turns of the spiral, as the collaboration increases, a community of the practice of collective learning develops. Etienne Wenger (1998) describes a community of practice as a group of workers in the same field who have developed ways of resolving the tensions between establishing common principles and breaking new ground. This is particularly important in dealing with transformational change where managing change is an inherent part of the practice.

When? Building a collaborative environment is important before the collective learning process begins, and for each of the steps of the cycle. While best face-to-face, sometimes online tools are a good option.

See: Six Revisions.
http://sixrevisions.com/tools/15-free-tools-for-web-based-collaboration/ [accessed 18.1.12]

Conflict resolution

Why? Diverse groups working together face possible conflict, especially when the groups are drawing on different sorts of evidence and experiences. Therefore a collective learning process needs a mode of conflict resolution that makes use of the conflict resolutions styles of all participants.

What? There are a range of conflict resolution methods and most people have a dominant style. In collective problem-solving the full range of styles is needed. Can you recognize the styles being used in a group and how to use them in keeping the group on track?

The Avoider: decides not to address the conflict directly. This might take the form of side-stepping an issue, postponing discussion, or withdrawing.

The Forcer: uses whatever form of power they have to force the response they want. Their power might be derived from experience, resources, rank, creative ideas or personality.

The Smoother: allows space for other's ideas and opinions. This may be generosity, giving in, or biding their time.

The Compromiser: looks for some mutually acceptable solution, through finding middle ground, trading concessions, or finding grounds for agreement.

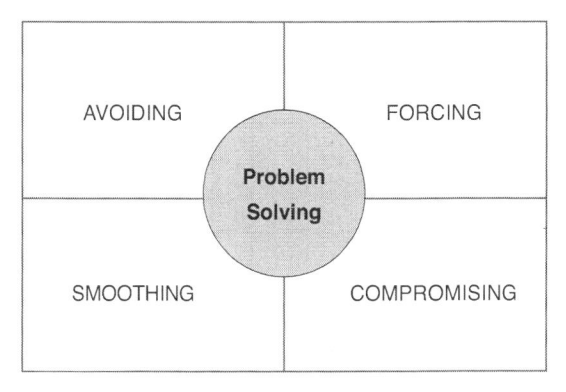

Figure C.1 The dynamics of conflict resolution

Personal strategies when faced with conflict in a group:

1. Establish rules for proper discussion.
2. Develop respectful responses to disrespectful behaviour.
3. Stick to issues and behaviours, not personalities or people.
4. Maintain your focus on, 'We can work this out.'
5. Count to 10. Use silence to increase your calm and cool the air.
6. Give people a way out. Establish choices.
7. Refuse the win–lose perspective.

When? Conflict is more likely to arise in Stage 2 of the cycle, describing 'the facts'. Different life experiences and ways of knowing will frequently generate different perceptions of what are 'the facts' surrounding a complex issue. Where conflict arises it should be acknowledged, and the differing perspectives each encouraged and accepted as valid in their own right.

Source of strategies: Local Sustainability Project (1990–2012).

Consultation

Why? Consultation in the form of dialogue between the participants is at the core of enabling collaboration, conflict resolution and all other aspects of collective learning. Unfortunately it is one of the most misused words in the lexicon of collaborative projects in general. Consultation is regularly used to describe passing on information of interest only to the sender, not to the receiver. A classic presentation of the degrees of consultation is that of Arnstein's ladder.

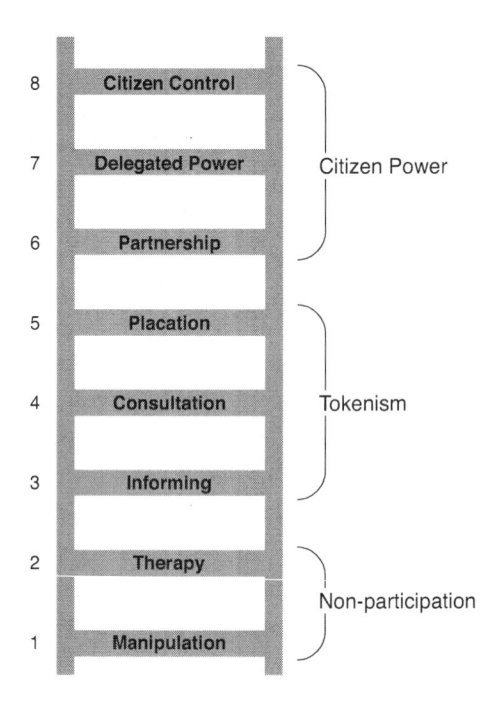

Figure C.2 Arnstein's ladder of public participation
Source: Arnstein (1969)

A standard mode of 'consultation' by many organizations, both government and industry, is to hold a public meeting in which the chief executive officer presents a formal account of their actions to anyone who chooses to come. If the meeting is held at short notice, is not widely publicized, or does not allow time for questions, the whole event becomes a travesty of communication.

What? In a collective learning project, consultation implies asking for information of importance to the exercise not already available from the participants. An important component of such a consultation is to be aware of the potential language gaps. The consultation may be between a collective learning team and an outsider from the community, government and industry or a specialist profession.

In each case the team will already be coming to understand each other, and the visitor will be speaking in their specialized language. Community members tend to talk in an in-language, referring to events and icons not familiar to outsiders. Government employees have a language all of their own, using mysterious acronyms (such as DEWAF – the Department of Environment, Water and Forests) and empty phrases ('We need to attempt to ensure that the escalating situation does not involve drawing on the legislative process' Is this equal to 'Better not to involve the law?').

Professional jargon is a by-word for being incomprehensible to outsiders. 'Cardiac arrest' for 'heart stopped' is a simple example. In this open communication era many departments and organizations employ special staff to act as translators from the in-language to the affected public. This is excellent when it works – AIDS (Acquired Immune Deficiency Syndrome) and SARS (Serious Acute Respiratory Syndrome) are examples of panic averted by transparent communication among colleagues and with at-risk communities.

A twist to this trend of engaging communication specialists to undertake the consultation is the temptation to bias the communication. One example of this is the 'spin' put on the information to sell a product or a policy. Another is the re-orientation of focus groups, historically a valuable means of consultation, to serve political ends. Yet another is push-polling, so called because a survey design includes questions which push the answers in a desired direction.

On a more positive note, there are directions the consultation needs to take when the collective learning team sincerely wish to capture opinions and ideas from another group. The questions that need to be answered before you begin are:

- What is the purpose of the consultative process? There will be different answers from those planning the consultation and those who are being consulted.
- Who needs to be consulted? Who are the people who are affected in any way by the issue as it is, or by the options for dealing with it, or by the way the options are implemented? These are the people whom you involve in the consultative process. As we discussed in Chapter 4, the term stakeholders can be misleading since it implies only those with a known stake should be consulted.
- What is the advantage to the people being consulted? Why should they give their time and energy to a consultation? There needs to be some hard evidence that the information gained will be put to some good use. It may be to inform policy, right a wrong, change course, or it may be some direct economic gain. People can be paid to come.

- What is the most suitable consultation process for this audience and this issue? This can best be decided through a working party composed of those consulting and those being consulted.

When? Consultation with other interests can extend the work of a collective learning team. Consultation is suitable to resolve a misunderstanding that has arisen, when not enough is known about how people would respond to a decision, when change is inevitable and people need to be informed. Tools are paper and phone surveys, face-to-face interviews, workshops, social media, door-knocking, snowballing, stalls in shopping malls, and special activities such as a world cafe.

In a world cafe, tables for four are set up, with one person designated as host. The rest of the participants join each table in turn for 15 minutes, a changing three at a time. All the hosts have the same set of questions that are answered by the series of participants. By the end of the time, everyone has had the chance to discuss and answer the questions, and will know what everyone else has answered. It is then easy for the facilitator to summarize the contributions, giving all equal value.

Also see: World Café
http://www.theworldcafecommunity.org [accessed 18.1.12]

Conversation

Why? However inspiring, no guidebook, strategic plan, leadership style, forecast or vision can bring the future into existence. The power to effect change rests with the people who are working together towards it. And that power rests with the conversations among the people committed to bringing about the change. Often the conversation goes 'How do we get others to work with us?' or 'How can we convince authority to support us?'. These are second order questions which only carry weight when the conversations between the committed group have developed a sense of identity, a sense of purpose, and a sense of power. To change the conversation is a springboard for transformational change.

What? Any gathering is an act of hospitality (see Hosting). In collective learning, everyone is host, everyone is accountable for the proceedings and the outcome of every meeting. The conversation among the participants gives the meetings their power. Such conversations carry commitment without bargaining, relatedness without obligations, dissent without argument. Conversations **not** to have:

- Telling and re-telling our own stories
- Blaming and complaining about ourselves or others
- Carefully defining fixed terms and conditions
- Seeking immediate action without negotiation
- Talking about people not in the room

The psychologist/engineer George Kelly observed that we define things by their opposites. This idea is useful as a basis for establishing the internal power of collective conversations:

- The servant creates the master
- The child creates the parent
- The citizen creates the leadership
- The student creates the teacher
- The future creates the present
- The audience creates the performance
- A room and a building are created by how they are occupied
- A shift in language creates a shift in behaviour

When? The effect of conversation can be immediate. Opening up a new conversation can change a culture. Transformational change is a change in the nature of things, not simply an improvement. A different kind of conversation is a vehicle through which transformation occurs. Without following the rules for dialogue as outlined in Chapter 2, it is all just talk, no matter how urgent the cause, how important the plan, how elegant the answer.

See: Collaboration, Consultation, Hosting, and Dialogue.
Also see: A Small Group
http://asmallgroup.ning.com/ Civic Engagement Booklet Series [accessed 22.1.12]

Dialogue

Why? Participants in a collective learning process bring with them their differing life experiences and ways of knowing. There is a need to establish a mutual understanding both of the words they use and the assumptions and reasoning behind them. It is from that shared understanding that new ideas emerge which allow transformational change. As physicist/philosopher David Bohm says, we learn from difference, not more of the same.

What? The Rules of Dialogue provide a valuable guide for all communications within any shared learning. Each step breaks new ground in human-to-human communication. A general adoption of dialogue is in itself a step towards a just and sustainable future.

Throughout the dialogue:

1. Commit yourself to the process
2. Listen and speak without judgement
3. Identify your own and others' assumptions
4. Acknowledge the other speaker and their ideas
5. Respect other speakers and give value to their opinions
6. Distinguish between inquiry and advocacy
7. Relax your need for any particular outcome
8. Listen to yourself and speak from the heart when moved to
9. Take it easy – go with the flow – enjoy

Source: Extracted from Bohm (1996)

When? Bohm's Rules of Dialogue are formally introduced at the beginning of any collective learning process, as in the introduction to the collective learning cycle. Dialogue is so different from the way in which we usually speak to each other that participants will need to be reminded of the rules. We typically reply to an opinion from others, with one of our own: 'My own opinion is ...'. In contrast, the first step in dialogue is to put your own case aside, and commit yourself to open-minded learning from what others think. Another way of thinking of it is to suspend disbelief, that is, accept that what other people are saying is true for them.

The second step, listening and speaking without making a judgement, is more difficult than it may seem. We readily make the reply 'I agree/ disagree…'. A dialogue-type response would be 'Have I understood what you are saying? The third step, clarifying assumptions, is crucial for you to be able to listen without making a judgement. It involves you being honest with yourself about what you are taking for granted. The fourth step is about indicating respect by making positive responses like 'How interesting', 'I didn't know that before'.

Steps 5 to 9 require a careful rethinking of how you usually act in groups. You will need to spend time thinking how you would put each one into practice.

Bohm (1996) gives more detail.

Event management

Why? Events and festivals can be vehicles for recruiting participants, project fund-raising and general advertising for any given learning cycle. Events have a large impact on their communities and, in some cases, the whole country. They provide an opportunity to send invitations to a celebration.

What? Event management involves identifying the focus audience, devising an event theme, planning the logistics and coordinating the technical aspects before actually executing the proposed event (see Case Study 2). Post-event debriefing is an important part of any event, including the collective learning process.

Event management is a strategic marketing and communication tool. From project launches to press conferences, promotional events communicate with the clients and potential clients of a transformational change. Possible events include press conferences, road shows, grand openings, concerts, award ceremonies, film premieres, launch/release parties, fashion shows and more personal events such as anniversaries and reunions.

In a large program, you can hire event managers to handle a specific scope of services for the given event, which may include all creative, technical and logistical elements of the event. The event manager is responsible for event design, audio-visuals, speech-writing, logistics, budgeting, and negotiation among the many people involved. Event management software provides event planners with tools to handle specific types of event. The key issues during the event are:

- Timetable
- Media contacts and press releases
- Health and safety including crowd management
- Amenities (access to toilets, food, exits)
- Speakers' needs
- Sound and lighting
- Detailed schedules (run sheets) for all key players
- Security

While event managers can provide very considerable assistance, the success of a collective learning event will usually depend in part on the key team members (the hosts) retaining strong links with the people and processes involved.

When? There are three points in a collective learning spiral when there may be a need to 'go public' with some sort of major event. One is at the beginning of a collective learning program when seeking a wide range of advice on the key issues and their proponents. The second is in planning for a collective learning workshop, where it is important to reach participants from all the decision-making interests. The third is at the close of each collective learning cycle when looking for major support for the action plans.

Also see: Kilkenny (2011)

Forecasting

Why? We live in an era in which the pace of change is greater than at any other time in human history. As a consequence every aspect of our lives is affected by change. While not trying to predict, those who have looked ahead will be best able to adapt to changes as they happen. Forecasting, often known as futuring, informs the stages of the collective learning cycle that look ahead: 'What could be?' and 'What can be?'. While visioning focuses on an imagined future (see Imagining and Visioning), forecasting seeks to identify probable futures and to plan strategically for them.

What? Futuring has been described as a process for standing in the future and looking back and back-casting as learning from the past. By exploring the big drivers of change and critically examining our own current assumptions and biases related to change, we can better prepare for and adapt to alternative futures. When people from different backgrounds come together to tackle wicked problems, the chances are that they will develop a robust set of scenarios for the future.

Tools that allow for developing probable alternative futures are brainstorming, computer conferencing, projections, trends, scenario writing, search conferences, modelling, cross-impact analysis, the Delphi process, the Nominal group process, role playing, simulation games and reading science fiction.

Care is needed to ensure that the future predicted is not the result of the tool being used. Remember 'if your only tool is a hammer, all the issues will look like nails'. Influential holistic accounts of possible futures have been Richard Adams' *Watership Down* an analogy with rabbits, *The Hitchhiker's Guide to the Galaxy* by Douglas Adams, and the classic *Brave New World* by Aldous Huxley.

When? Forecasting is approached differently at each of the six steps of a collective learning cycle. In scene-setting, forecasting can provide the basis for a well-informed focus question. On the other hand, forecasting should not be used to answer the first question, 'What should be?', since this step is for an ideal possible future, not a forecast probable one. Forecasting provides an excellent means of deciding 'the facts' for a probable future and it can stimulate and ground new ideas in learning stage three. In learning stage four, 'What can be?', futuring is a useful tool for grounding the action plans. At the sixth step of each cycle, following on, it can be used to predict the conditions under which the next step of the cycle will operate.

Also see: Makridakis (1997)

Gatekeepers

Why? A collective learning cycle aims to maximize the benefits of participants' diverse knowledge and experience. Gatekeepers are the people who control the information flow. Negative gatekeepers close the gate. Positive gatekeepers keep the gate open to all comers. As a positive gatekeeper they can also choose to close the gate to people who attempt to dominate the discussion.

What? Positive gatekeepers are strongly group oriented and have good external networks. They keep open the channels for information flow within, and in and out of the group. Gatekeepers often act as facilitators and negotiators, and can be found in most social roles. Taxi-drivers and bar-keepers, nurses and community workers, managers and tea-ladies have been documented as opening up communication channels for people who have trouble finding their voice.

There are various ways in which a positive gatekeeper might operate in assisting a transformational change. Some may act as jokers, like the court jester of old. Introducing political cartoons and entertainers such as the Beatles during the learning cycle can initiate changes in thinking. On the other hand, a joker can distract the dialogue from the business at hand. Other gatekeepers can act as organizational termites: 'eating away' at fixed ideas and limited interactions behind the scenes while building the pathways for new voices to be heard.

Another type of gatekeeper is the positive deviant: someone who acts differently from the socially approved behaviour, and thereby acts as a role model for change. An example is a mixed sporting team during the South African apartheid. In any organization there is a nexor, someone everybody talks to, rather like the old meaning of gossip.

Any of these positive gatekeeping roles can be played by anyone from any of the interest groups, from the chief executive officer of an organization to a community bystander. In the design of a collective learning cycle, positive gatekeepers can be identified and used to good effect.

When? Gatekeepers can emerge at any point during a collective learning cycle, or they can be deliberately sought as part of the process. Some may adopt their role as the cycle progresses. Others enter the process with a positive participatory approach and become positive gatekeepers when they find all views are not being heard or addressed.

Also see: Coffee (2009)

Hosting

Why? Any party that brings together people from many different backgrounds relies on the combination of careful planning and the spontaneous warmth of the host. In much the same way, collective learning benefits from having the leader or the leadership team act as welcoming hosts.

Dictionary definition: A host invites guests to a social event, such as a party in their own home, and is responsible for them while they are there. This extends to the host of a television show or a hotel. These definitions don't convey the nature of a host: to be warm, welcoming, convivial, friendly, helpful and able to be relied on.

What? The role of 'host' in a collective learning process is more active than a facilitator, and less formal than a leader. A host is there to help people feel welcome and comfortable, guide the process and keep it open to all involved.

Any good host will:

- Offer a warm welcome
- Arrange a suitable setting
- Try to meet their guests' needs
- Provide food for their stomachs, their eyes and their ears
- Offer some form of entertainment
- Encourage conversation
- See their guests off with a warm farewell

All of these can be translated into a collective learning event. In each of the case studies there is an opportunity to get to know the guests beforehand (setting the scene), clarify their needs (focus question), include a convivial meal whenever possible, show visuals as well as talking about the project, break up the learning stages with activities drawn from the toolkit, leave the guests to do the talking, and finish with a happy feeling for them to remember.

When? Like any good party, the collective learning host will be active from when the party is planned and the invitations issued, until the last guest has left and the follow-up is in train. Each type of celebration asks for a different format, and guest list. A bon voyage party is different to a housewarming, and different again from a bring-a-plate gathering.

Also see: Art of Hosting. http://www.artofhosting.org/home/ [accessed 18.1.12]

Imagining

Why? Imagining a new future in a just and sustainable world is something that many people around the globe are finding hard to do. Imagining a future where collaboration has replaced competition, that understands the boundaries between human and natural systems, and that provides a fair deal for everyone in the community presents many challenges to us in our everyday world.

Imagination is associated with creativity, insight, vision and originality. It is also related to memory, perception and invention. All of these are necessary in addressing the uncertainty associated with wicked problems in cases of transformational change.

What? The challenge is to enable a mix of people to generate 'blue sky' options. When using a collective learning approach, it is important to begin with 'What should be' (the ideal), rather than what is now, thus freeing participants from an initial burden of current constraints and challenges. Providing a non-judgemental climate in which there are no 'rights' and 'wrongs' or good and bad, and encouraging participants to think outside the square, are also important to creative imagining. Even the zaniest ideas may stimulate new connections and the new thinking that is needed.

The use of visual rather than verbal tools often frees participants of their culturally derived boundaries and 'permits' them to use the creative right side of the brain. This allows access to deep feelings for which there may not be words. Having the participants draw a picture of the issue or of its solution is often a useful tool. Sharing the images is an important step in collective learning. Case Studies 2, 10 and 15 have examples of participants linking visual images into a single shared picture.

'Guided imagery' as used in health sciences and psychology involves meditation exercises that help participants dream of a different world. The imagery may be a journey, a Shangri-la, or a magic carpet that takes them into unknown territory. More pragmatically, the individuals in a group can take an object, such as a car, and label the parts of a future world. Who is steering? Who or what acts as the wheels and the passengers, the petrol, the windows, the brakes?

When? Setting the right conditions for using the imagination is particularly useful at the first learning stage of the collective learning cycle, when participants come together to define 'What should be' (i.e. to develop a shared vision). Imagining may again prove useful when the group moves to Stage 3 in the collective learning cycle – identifying 'What could be'.

Also see: Koestler (1990).

Joining in

Why? The success of a collective learning process is critically dependent upon the active involvement of people from different backgrounds and life experiences. However, there are many factors that prevent people from joining in. The reasons may be shyness, caution, fear, feeling strange, or dislike of the leader or fellow participants. They may be tired, sick, grieving, or overworked.

For everyone in a diverse group to like each other or agree with each other is unrealistic. It is even unhelpful where conflicts need to be aired. To join in does mean giving up some degree of independence. Therefore the collective learning process needs conditions that allow people to feel safe and able to trust one another, no matter how different their views are.

What? The way in which participants are recruited will have a marked influence on how individuals join in. The initial invitation is crucial. It needs to tell participants they will find value for them, and present them with opportunities to influence the outcomes.

A disability consultation held in an upstairs venue without lifts, a pre-school planning session held mid-afternoon when young parents are collecting their children, a regional fire management planning workshop held at the peak of the rural fire season, and an Indigenous session soon after someone has died are unlikely to encourage people to join in or to respect those who have organized such sessions.

It is important to consider the location, time and format of any shared sessions if we want them to be inclusive. For many participants, travel time and costs may need to be reimbursed. Once in the room, participants in a collective learning process need time to introduce themselves and/or to say what is their interest in the issue. This might be done in a round robin or individuals may be asked to find someone that they don't know well, do the introductions one-on-one, then report back to the rest of the room on what they have learnt about each other.

When? Ensuring that the right people join in from the beginning is important to the success of a collective learning process. Even once a collective learning process is under way, it is important to retain some flexibility to accommodate changes as trust grows, shared understandings develop, or the unexpected emerges from the dialogue.

Also see: Ledwith and Springett (2009) and Skolimowski (1995).

Knowledge brokering

Why? To learn is to add to one's knowledge. This may be knowledge of how (to do something) or knowledge that such-and-such is the case. Everyone learns all the time simply through being alive. In accessing significant new knowledge, it is helpful to have a knowledge broker – someone who acts as a bridge between the learner and the new knowledge. In formal education this role is taken by a teacher. In a collective learning cycle, everyone in the group acts as a knowledge broker for everyone else, helping them see familiar situations through fresh eyes.

At each of the four learning stages of the collective learning cycle, an external knowledge broker can act as facilitator, translator, negotiator, researcher and scene setter, according to need. While many decisions are made automatically without need for brokering, transformational change brings with it a need to make major decisions under new conditions, and this often needs a knowledge broker.

What? What is knowledge anyway? We all know more than we think we know. As well as explicit knowledge, there is implicit knowledge (what we don't know we know) and ignorance (what we don't know we don't know). The dictionary says knowledge is justified true belief. In other words we know something when it has been checked out against some test for truth. This may be our personal experience (individual knowledge); what everyone around us believes (community knowledge); something that has been observed and measured (specialized knowledge); ways to achieve a group goal (organizational knowledge); or a sense of fit with all the evidence (holistic knowledge). Each of these gives us a different way of validating the 'truth'. Then we believe it, and call it knowledge.

Material transmitted through libraries, lectures, communication media and gossip is information until it is absorbed into the mind of the learner. In Western society each individual learns to prefer one source of 'truth' over another, so there is need for a broker so that each group can learn how to share the knowledge of others.

When? Think of committee meetings, community projects, and specialist seminars in which everyone argues. With the help of a knowledge broker, the participants could equally well use their time to develop a constructive synergy.

Also see: Lomas (2007).

Learning styles

Why? Learning in children is constant, guided along certain paths by a combination of genetic conditioning and social environment. In adults, a great deal of learning has already gone into making them who they are, and they now learn as distinct individuals. This means a call for fresh learning can threaten what they already know and how they have learnt it. So in any process of collective learning, there is a need to take account of the different approaches to learning among the group, or they cannot learn from each other – and sometimes are blocked from learning anything at all.

What? Chapter 10 contains a list of possible learning styles which need to be taken into account in collective learning. David Kolb used a word test to identify the learning styles of over 5,000 adults (Kolb et al. 1986). He found that the participants in the study fell spontaneously into four learning styles – and that a learning style matched certain occupations and approaches to addressing complexity. Accommodators welcome and build on complexity; assimilators seek to reduce complexity to order before they can learn. Divergers address complexity by looking outside the square; and convergers think linearly and look within the existing evidence. In a group of mixed interests, all or any of these learning styles may be present, and individuals may switch from one to the other as circumstances change.

Learning styles, problem-solving, and methods of inquiry are closely related. Thus Edward de Bono's six thinking hats for everyday problem-solving (1999) can be regarded as a popularisation of David Kolb's learning styles and of the range of inquiry methods in multiple decision-making. They also overlap with Howard Gardner's multiple intelligences (1983). Each of the lists makes it clear that it is important to identify the context in which the learning is taking place, and which format the participants will find most useful.

	De Bono problem-solving	**Gardner** intelligences	**Kolb** adult learning	**Multiple knowledges** collective learning
White hat	Information	Logic	Convergers	Specialists
Red hat	Feelings	Intra	Accommodators	Community interpersonal
Black hat	Judgements		Assimilators	Organizations
Yellow hat	Optimism			
Green hat	Creative	Creative	Divergers	Holists
Blue hat	Thinking		Convergers	Specialists

When? It is essential that the different learning styles of a particular collective learning group are considered in designing a collective learning process and then as the process proceeds. Many of the tools can be used to diagnose the styles either before or after a group has formed. The styles above need to be considered at each of the steps of the learning cycle in Chapters 3–9.

The sources are referenced in the bibliography.

Multiple knowledges

Why? The complex problems that society faces are generated by our modern society and thus require new ways to address them. Climate change, food security, sustainability and a host of other complex problems facing our world are being labelled 'wicked problems' – problems that demand fresh thinking in the society that caused them (see: Chapter 10 and Wicked problems).

It is therefore important that we draw on all the wisdom, knowledge and creative thinking we can to bring about the transformational changes needed to address these problems. That means combining multiple knowledges.

What? Chapter 3 describes the five knowledge bases needed to effect whole-of-community change from the point of view of who to invite into the collective learning circle. Research has confirmed that any transformational change needs the knowledge that key individuals, affected communities, relevant specialists, influential organizations and creative thinkers bring to the process.

Combining all the knowledges presents a challenge to the ways we make decisions at present. The usual process is to seek a consensus, calculate an average, or find one right answer. In addressing complex problems, we need a collective answer that respects all contributions. There is no single magic bullet. Consensus without collective learning simply reproduces what is, and does not open up the possibility of transformational change. The mandala in Chapter 4 is a symbol for combining multiple knowledges, although it is never as tidy as this figure represents it.

When? These multiple knowledges have to be taken into account in all collective learning. When the learning is how to facilitate transformation, then the five knowledges are essential. Transformational change requires the support of all those who have to live with the change. Everyone can access all the knowledges for themselves, although we are trained to concentrate on one at a time.

Also see: Brown (2008).

Negotiation

Why? Negotiation is needed whenever different individuals come together to work on the same issue from their different points of view. Negotiation is about respecting the diversity of the people involved and their values. Given the perspectives and interests that are involved in any transformational change, negotiation is needed at many points along the way.

What? Successful negotiation is critically dependent upon the participants coming to the table with goodwill and a commitment to find shared ways forward. It is up to the facilitator to establish this learning environment.

In any negotiation it is important at the outset to agree on the central question that everyone is seeking to address – the focus question that provides all participants with an agreed point of reference. Collective learning seeks to go beyond 'win–win' solutions, to a new perspective, and hopefully a collective outcome.

Fisher and his colleagues in *Getting to Yes* (2011) highlighted the value of 'principled negotiation'. Rather than adopting a 'positional bargaining' approach in which each party in the negotiation pits their will against that of the others, the participants in a principled negotiation come together to seek to reach a 'wise agreement amicably and efficiently'. Fisher discusses the four key principles that underpin principled negotiation:

- Separate the people from the problem (and focus on the problem)
- Focus on interests, not positions
- Generate a variety of options/possibilities before deciding what to do
- Insist that the result is based on some objective standard or criterion

It is important to remember that in a negotiation process, there are no 'rights' or 'wrongs', rather there is a conscious, shared commitment to finding a new way forward for all involved. A skilled and impartial facilitator with some knowledge of the subject matter is needed when there has been a history of conflict.

When? Principled negotiation is useful throughout the collective learning cycle, but is perhaps most needed at Stage 4 when participants are seeking to agree what actions not only should but can be taken and who will be responsible for each action.

See also: Fisher et al. (2011)

Open Space Technology

Why? Collective learning asks for an Open Space where all those in the learning group can share their knowledge. Sharing knowledge is much more intense than simply exchanging information. Facilitator Harrison Owen (2008) developed Open Space Technology to facilitate dialogue (see Dialogue) that addresses an issue that is important to all the participants.

What? Open Space Technology is an appreciative or strength-based approach. It assumes that every participant has something to offer the group and that passionate and committed people will be willing to work together in accordance with four principles:

- Whoever comes are the right people
- Whenever it starts is the right time
- Whatever happens is the only thing that could have happened
- When it's over, it's over (i.e. do what you feel is of value, then when that's done, move on)

An Open Space process begins by an appointed facilitator making sure everyone understands the rules and shares a common problem. The participants are then invited to paste their priority topic for discussion on a bulletin board. If a group forms around that topic, the proposer takes on the responsibility of leading the group in a dialogue so that everyone can be heard. Each group develops an agenda, and produces a program of action.

When? Open Space rules are basic rules for every part of the collective learning cycle. The learning stages are constructed on open space principles. Each stage is a blank space until the participants share their knowledges and spark off a synergy from which everyone learns even more. When a set of groups come together on a shared issue it is all the more important to construct a transparent agenda which everyone in the group shares.

Those who are used to structured meetings and workshops in which the content is determined before the event may be uncomfortable with an Open Space process. A level of trust in the process, the facilitator, or both is needed since participants are being asked to take shared responsibility for the agenda, the themes addressed and the outcomes reached.

You can find more about Open Space Technology at www.openspaceworld.com.

Pattern languages

Why? In guiding transformational change, the interactions between interests are complex and at times confusing. A shared language is often the best way to describe an issue and generate the best possible collaborative solution. Developed by architect Christopher Alexander (2002) for use in community-based urban design, pattern language has spread to design fields such as computer software, engineering, social media, civil action, administration and business. It is therefore already an avenue for bringing together very different problem-solving approaches.

What? A pattern is an interconnected word picture of the whole of an issue. Each pattern describes the context of the issue, the core problem to be resolved, the forces impacting on the issue, a potential solution and examples of the solution in practice. Alexander identifies a good pattern as one which brings life to the issue and has a strong sense of meaning.

EXAMPLE
A pattern language for developing civic intelligence

Context: Humanity and the natural environment face profound changes that will require an immense amount of attention if the challenges are to be met. Society needs ways to develop intelligent collective responses that include the voices of the citizens.

Issue: In transformational change the whole community is involved in the change. The citizens need ways to develop their civic intelligence, that is, a flexible and powerful competence to deal with the community's own issues. This goes far beyond the traditional narrow notion of intelligence.

Forces involved: The power hierarchy among knowledges (see Chapter 10) means that citizens' voices often go unheard. Democratic processes are often not enough. On the other hand, a community has power in its own right when it chooses to exert it. A community does not need to be given power, but it does need non-violent ways to put that power into practice.

Solution: Civic intelligence combines the contributions of many individuals and focuses in creative ways on having their voices heard.

Practice: Citizen action events include street marches (anti-nuclear demonstrations, gay parades), use of social media such as Twitter (Iranian protest movement), distributing posters (pictures of heroes, examples of injustice), non-violent action (the freedom riders in segregated buses in southern USA and Gandhi's Indian salt marches).

When? Describing an issue in the form of a common framework such as a pattern can be a useful way of arriving at effective collaborative action.

Also see: Schuler (2006)

Problem-solving games

Why? Often members of a group have experience only in their own traditional ways of problem-solving. Different members may expect to go to an authority, access a search engine, ask their friends, go it alone, look for a recipe, take authority themselves or give up when it's hard (see Xing the minefield for an example). Simulation games allow the group to learn their own and each other's problem-solving approaches and to find effective approaches that they can share.

What? Problem-solving games can be designed for almost any type of issue. From mock auctions, members can learn what it means to go to any lengths to win and what it means to give up too early. The prisoner's dilemma is a classic version of losing when you don't share, winning when you do, and how hard it is to choose the second.

Building a tower together as a team, from whatever materials are provided, can be an exciting simple way to identify leaders and followers, innovators and caution. Given the challenge, it also allows the group to practise being a team.

Communication games include Chinese whispers – sending a message around a circle and noticing how it becomes twisted. Another is to sit in pairs back to back and describe an object only one can see. Yet another is for each to draw the situation that brings them together and then form a single picture (see Case Study 2).

Trust-building games include leading each other blindfold through familiar territory, each member taking a turn falling for the team to catch, and risk-taking such as high ropes courses where each of the team is dependent on the others for safety.

Ice-breakers allow group members to get to know each other quickly, and at greater depth than simple introductions. In 'The dance card' people fill in a card through finding people with shared interests: sport, relaxation, work, family etc. 'The value line' is described in Case Study 13. Members can introduce each other in pairs, or in trios that find things in common and differences.

When? Problem-solving games are particularly useful when participants from different backgrounds come together for a collective learning process.

A warning note: each of these games can be trivialized and left without any significant learning. On the other hand, they can open up depths of feeling and relationships that people do not wish to share. In either case, the facilitator/guide is responsible for preventing either of these undesirable outcomes, and for making the game a valuable and enjoyable enterprise. Sensitivity and practice are required to design and facilitate an effective game that brings with it the desired positive learning.

More games at: http://inspireyourgroup.com/blog/activities/group-problem-solving-teambuilding/ [accessed 18.1.12]

Questioning

Why? One of the principles underpinning a collective process is the need for all relationships to be transparent and equitable. Appropriately used, strategic questioning can catalyse the exchange of ideas that stimulate new thinking around a complex problem.

What? Closed questions are those that require a simple (usually single word) answer, such as 'Yes' or 'No'. 'Do you think this issue is important?' can sometimes be useful but adds nothing to mutual understanding. The open question 'Why do you think this issue is important?' provides more information. Open questions require a more detailed and considered answer. Questions such as 'Tell me what you think of this issue?' will take longer to respond to and will likely require more probing questions to get the full picture. Strategic questioning is 'the skill of asking questions that will make a difference'. Practitioner Fran Peavey designed questions that reveal what she called 'Heart politics' which allow a greater depth of sharing.

A strategic question creates movement: 'Why did you move to... ?', rather than 'Where did you move to?'

When? There are many opportunities for using strategic questions in collective learning: as a focus question in providing focus both at the beginning and during a collective learning cycle; as the trigger for the visioning questions in Stage 1 of the cycle; to help explore obstacles and consequences at the 'What is' fact-finding Stage 2 in a cycle; and to encourage participants to consider all the alternatives and to think about how change can best be achieved (during Stage 3 of a collective learning cycle).

A valuable resource is Fran Peavey's *Strategic Questioning Manual*, available online at www.thechangeagency.org/_dbase_upl/strat_questioning_man.pdf [accessed 18.1.12]

Risk and risk-taking

Why? Risk is an ever-present part of life in our complex and uncertain world. Our perceptions of risk influence our decision-making and our behaviour, but people from different knowledge cultures will likely have different perceptions of risk. Since risk and an element of risk-taking (going beyond the known and exploring new ways of doing things) are essential to the transformational changes needed to address complex and wicked problems, it is important that we understand the ways in which different people perceive different risks.

What? Risk can be perceived in one of four different ways: calculable risk attributable to human actions and predictable in quantitative ways; danger, which is also attributable to human action but which is not readily quantifiable; fate and uncertainty, which are unexpected 'acts of God', not attributable to human action; and omnipresent risk, which although attributable to human actions is ever-present and unquantifiable.

Both our temperament and our life experiences and training influence our perceptions of risk. While engineers are trained to be risk-averse and to make their systems as fail-safe as possible, ecologists working with natural systems are trained to place much more emphasis on adaptive management (learning by doing, evaluating and reassessing).

Those trained in physical sciences and economics will likely place greater emphasis on calculable physical risks that are more predictable, while those trained in social sciences will likely focus more on intuitive and less certain social risks.

Sandman (1986) identifies risk as the sum of physical hazard (as perceived by specialists with technical knowledge) and outrage (the fear, dread or misery perceived by people in a community). According to Sandman: Risk = Hazard + Outrage and all parts of the equation need to be addressed in building shared acceptance.

In any collective learning process in which people from diverse backgrounds and knowledge cultures come together to address complex or wicked problems an awareness of the social aspects of risk perception and activities to address those perceptions are important. Our perceptions of risk influence not only how we respond to each other in a collective learning situation, but also how we structure our engagement with the wider community.

The level of risk associated with an issue, the complexity of the issue and the information that relates to it will all influence the level of participation required by the community. These same factors will also lend weight to the investment of resources (both financial and human) that is required.

When? Going beyond what is known and into a world of uncertainty is essential if we are to achieve the transformational change needed to address wicked problems of modern society. This is particularly important at Stages 1 ('What should be': defining the ideals) and 3 ('What could be': new ideas from collective learning) of the collective learning spiral. However, participants in a collective learning process will need support, encouragement and a sense of trust in their fellow participants to step beyond what is known and 'safe'.

For a theoretical overview of risk, see Adam et al. (2000)

Synergy and synthesis

Why? In addressing complex or wicked problems, it is necessary to bring together the rich mix of knowledge that exists within different cultures and life experiences. Synergistic inquiry exists in a space in which relationships and interactions enhance the likelihood of synthesizing a new whole that goes far beyond any of the individual ideas brought to the table by participants.

What? Synthesis is the creation of a fresh new 'whole' formed from the interacting parts, while synergy is the interaction between the parts in ways that enhance both the parts and the whole. Synergy is not a jigsaw puzzle. It is a collage. In a collective learning process, participants are striving for a new whole that goes beyond the consensus outcome that might be achieved in a traditional decision-making process.

We live in an era in which many of the problems faced by communities, whether local or global, are wicked problems requiring transformational change. This is an era in which we need a collective learning situation rather than the separate analytical approaches of specialist technical disciplines.

Although 'facts' are expected to define reality, people from different sectors will see those 'facts' very differently. In a collective learning process, all need to be identified and acknowledged as a basis from which to move ahead.

Australian Landcare re-shaped farming practice from exploitation of the land towards partnership with the land, even though farmers and conservationists had been at daggers drawn for decades.

When? To achieve synergy, we need to set aside the usual competitive approach to learning that has evolved in current Western culture and to replace it with a collaborative effort. While synergy and synthesis of ideas and knowledge are important throughout the collective learning cycle, they are perhaps most important at Stages 2 (defining 'What is'), 3 (What could be': creating new ideas) and 4 ('What can be': shared action).

For a detailed discussion of synthesis and synergy see Brown (2008).

Team-building

Why? A team can be thought of as a group of interdependent individuals who share a common purpose, and take responsibility for whatever happens among them. Team-building becomes especially important in collective learning where a diverse group is brought together to address a wicked problem.

What? Within any given team there will be people with different values and priorities, people who use different language to describe the same situation, and people who, because of their temperament and past experience, prefer to take on different roles in the team.

Think of the economist, the environmentalist, the engineer, the scientific researcher and the long-time landholder who come to a project as community representatives. The different levels of distrust, based on past experience, mean that the problems for achieving transformational change are magnified.

A skilled facilitator can serve an important role as a 'cultural translator' making each of these positions accessible to the whole group. They can help group members seek out the roles that they can perform most effectively. These roles may include: coordinator, slave driver, finisher, critic, ideas person, resource-finder, and supporter (and see Conflict resolution and Risk-taking roles).

People need to know each other as individuals, not just as scientists, community members, or representatives of organizations. Tools that can be used to help build that mutual understanding include:

- sharing rituals such as a special greeting or bringing a special cake
- story-telling
- sharing things about themselves and their lives
- identifying each member's favourite animal, song, sport etc.
- choosing from among the many ice-breaking exercises e.g. set up a value line, list ten favourite books to read in the bath, decide who goes in a life-boat
- introducing activities that require cooperation and trust
    - one member of a group navigates a course blind-folded using clues provided by the others
    - asking all participants to identify what for them personally are the high-risk actions they might have trouble with (e.g. asking help from others, expressing a difference in opinion, admitting 'I was wrong... ')
    - an 'appreciation exercise' – 'what I value about you is... ' which encourages people to feel good about themselves and their colleagues.

As people come together in new collective learning teams, they need to be aware of the predictable stages in team development. These stages are:

- **Forming:** getting to know you. Team goals and objectives are not yet fully defined, roles and responsibilities not yet clear, and members stick to 'safe' patterns of behaviour.
- **Storming:** who is boss here? As tasks become defined, members vie for positions, and tensions arise from their differences. To move on, group members need to share tasks that allow them to know each other better, as in working through the stages of the collective learning cycle.
- **Norming:** getting on together. Patterns of agreement form among team members, individual contributions are being acknowledged and decisions are made collectively.
- **Performing:** getting on with the job. The team has a shared goal and agreed tasks which are being achieved in interdependent ways.

- **Mourning/reforming:** moving-on. The team winds up with regret and tries to find ways to continue collective tasks and relationships.

When? Team-building is essential throughout any collective learning process. It should be introduced in ways that are non-threatening to those who are less confident or trusting.

Also see: Steinberg (2004).

Transdisciplinarity

Why? The task of collective learning is to find ways to draw on all our intellectual resources, valuing the contributions of all the academic disciplines as well as other ways in which we construct our knowledge. And that brings the challenge of developing open transdisciplinary modes of inquiry capable of meeting the needs of the individual, the community, the specialist traditions and influential organizations and allows for a holistic leap of the imagination (see multiple knowledges).

What? 'Transdisciplinary' is taken here to be the collective understanding of an issue; it is created by including the personal, the local and the strategic as well as specialized contributions to knowledge. This use needs to be distinguished from a multi-disciplinary inquiry, which is taken to be a combination of specializations for a particular purpose, such as in a public health initiative, and from interdisciplinary, the common ground between two specializations that may develop into a discipline of its own, as it has in biochemistry. 'Open' transdisciplinarity includes the disciplines, but goes further than multi-disciplinarity to include all validated constructions of knowledge and their world views and methods of inquiry.

Transdisciplinarity in the broad sense requires the use of imagination by all concerned, since it brings together different ways of thought that have not informed one another before. Imagination is associated with creativity, insight, vision, and originality; and is also related to memory, perception and invention. All of these are necessary in addressing the lead-up to and the consequences of transformational change. In a practical sense, imagination has been central to the work of anyone who is involved in changing the society in which they live. Accepting a central role for the imagination does not mean that we abandon standards for assessing the validity and reliability of the knowledge so generated; it indicates the potential for change and shows us where to look.

Transdisciplinarity tackles complexity in science and challenges knowledge fragmentation. It deals with research problems and organizations that are defined from complex and heterogeneous domains such as climate change or housing and health. Transdisciplinary inquiry accepts local contexts and uncertainty; it is a context-specific negotiation of knowledge. Transdisciplinarity implies continual interactive communication among the diversity of interest groups. Transdisciplinary research and practice require close and continuous collaboration.

See Imagining, Multiple knowledges and Wicked problems.

Also see: Brown et al. (2010).

Understanding

Why? During transformational change, there are many opportunities for participants to misunderstand each other. Mutual understanding depends on the use of dialogue and keeping minds open to the different and the new.

What? Differences in our values and ways of knowing are often reflected in the words we use to describe something, and this in turn can lead to different understandings of the same event. For example, in rural Australia a group of farmers and ecologists used the following terms to mean the same things:

Community	Ecologists
The balance in nature	Biological diversity
A patch of bush	Remnant vegetation
Healthy country	Resilient landscapes
Grass and scrub	Understorey

Since it is important to ensure that everyone understands each other, it can be useful to introduce tools that allow for mutual translation. Some of the more useful tools include:

- Drawing rich pictures
- Values mapping
- Sharing individual understanding of the same task
- Walking-and-talking sessions in the field

A facilitator can do much to help participants understand each other – avoiding jargon, seeking clarifications, and paraphrasing.

When? Shared understanding is important throughout a collective learning process, from the initiation of the process, through invitations and the four stages of collective learning to the final follow-up.

Also see: Brown (2011).

Values mapping

Why? Each of us brings with us a set of values that guide who we are and how we behave. Our values are deeply embedded and have a significant influence on our priorities. Differences in values lie at the heart of many disputes and conflicts, especially when people from different backgrounds and life experiences come together, as in a collective learning process.

What? Values mapping is a strategy originating in business to assist companies to define their own direction and to position themselves in the marketplace. However, by combining the visual mapping of values with processes used in conflict resolution, Judy Lambert and Jane Elix have developed values mapping as a tool to assist participants in collective learning projects moving towards transformational change.

The use of a map of the focus area enables participants to move away from the language that entrenches fixed positions and to share each other's perspectives in a less threatening way. In this process all values are valid and it enables a sharing of perspectives from which alternatives might be generated.

The essential steps in this consensus-building approach are:

- Background research through which the facilitators better understand the issues, the language used and the history to the issue
- Identifying the key players, their roles and interests, and their dominant ways of knowing
- Designing the mapping process in ways that are accessible and meaningful to all participants
- Establishing 'buy-in' through the encouragement of participation and building of trust
- Mapping values through focus groups, site visits and consensus-building workshops
- Reporting back to all participants

When well defined, values act as a point of definition that guides the focus of a changing and dispersed workforce and enables high involvement and rapid response. Values mapping is complemented by value positioning, which Involves a three-step process – diagnosing (in collective learning identifying 'What should be'), identifying ('What is') and repositioning (to 'What could be').

When? Values mapping is used at various stages in the collective learning cycle. Where groups are at odds over the issues to be addressed in achieving sustainable management outcomes, values mapping can be used to assist in developing a shared vision for the future ('What should be'). Once that vision is in place, the values map can also be used to identify the critical issues – defining the 'What is'.

For more information see Elix and Lambert (2007).

Visioning

Why? Sharing each other's visions is an important starting point for giving a direction to collective learning and action for a transformational change. Visioning exercises also energize commitment from the participants.

What? A diverse group of individuals is asked by a facilitator to imagine themselves in a situation 15 or 20 years hence and to describe how things are and what they see in their ideal world. In a vision there can be no 'right' and 'wrong' views – all are important and it is often the sharing of diverse ideals that allows collective learning.

A short version of a script often used in a visioning exercise is:

Aim: To establish a shared holistic vision of their collective future for local citizens, community interests, local specialist advisors and responsible government agencies.

Facilitator: Ask people to relax, take off shoes, tight belt etc. and close their eyes. Speak slowly and quietly. I want you to come with me into our helicopter and first take it up high over **[Place name]*, not **[Place name] now, but in 20 years' time when all our effort toward a better future has succeeded. Look at the landscape beneath you. What does it look like? What colours and shapes do you see? What time of year is it? What patterns do you see? Where are the people? Who is in a family with whom? Can you see people and goods moving to and fro? How are they being moved? What sounds come up to you? What smells?

Descend slowly, looking around you as you come down at the patterns of buildings, of people, of spaces. Look at the shapes and structures of the buildings. Look at the spaces between them. Who is moving in those spaces? As you get closer, listen to the sounds and smell the scents of the landscape.

Are there people around? Who is there? What ages are they? How do they react to you – and to each other? What activities are going on? Are there people working? Are there people relaxing? What else are they doing? Who is doing it? Now walk around and see what else there is.

Join some of the people as they move about. What are they doing? How do they look? Is this a good place to be? How are the people interacting with each other? Now join some other people. What is happening here and how does it feel?

Now imagine yourself in the open spaces. How is it different from today? What is happening in the neighbourhood and in the community? What are people happy about? What are they worried about? Imagine that it is lunchtime.

Where do people go for lunch? What do they talk about? What sort of food are they eating? Where did it come from? What do they do after lunch?

As the day comes to an end, go home with someone. How do they get home? Where is home? How far are they travelling? Go with them in their neighbourhood. Remember, this is a well-managed environment in which all your programs and changes are working. How does the neighbourhood feel? Do you feel secure about the future? What important changes have happened recently? Now go into a house. Who is there? Is it a family? Who is in the family? Are they all receiving an income? What do they do? What about the others? Are there children? Do they go to school?

Slowly come back to the present time, thinking about all you have seen. [Provide time here to return to the here and now]. Now, before you forget what it was like, write down the 10 most noticeable things you saw, heard, smelled or touched.

Give the participants five minutes or so to write down the 10 things that stood out for them. If necessary, put some pressure on the participants to go to 10 items – the last two or three may be the most difficult, but often contain the most original ideas. Next, ask people to rank their 10 ideas or issues in order of importance to them.

The group members share their list in order, the top one first, then the second and so on. Stick-on notes on a whiteboard, taken in turn with each person reading out their own and answering questions of clarification, are good ways to share. The whole group then looks for shared experiences and themes. If desired, the themes can set the change agenda.

When? Visioning is a very powerful way to help a group access their ideals in Stage 1 of the learning cycle. There will always be someone who is very practical, or not at all visual, who has trouble doing the exercise; that's fine. They should be made to feel comfortable and asked to complete the 10 ideas from guesswork or their own experience rather than guided imagery.

See: Brown (2008).
Also see: Vancouver Community Network. http://www.vcn.bc.ca/citizens-handbook/2_16 visioning.html [accessed 18.1.12]

Wicked problems

Why? A wicked problem is in a class of complex problems that cannot be resolved without changing the society that generated it. On the global scale examples are climate change, urban violence and the obesity epidemic. On the local scale it may be youth drug use, industrial accidents, or extinction of a species. In any case, the solution does not lie in more of the same but in transformational change.

Two management experts, Rittel and Webber (1973), labelled these problems wicked, because they resist standard ways of problem-solving. A more detailed account of wicked problems can be found in Chapter 10.

What? While no wicked problem is exactly like another, they have some common characteristics:

- Wicked problems evade clear definition. They have multiple interpretations from multiple interests, and so no one version can be considered right or wrong.
- Wicked problems are multi-causal with many interdependencies, therefore their resolution involves negotiation between conflicting goals.
- Attempts to address wicked problems will lead to social change and so to unforeseen consequences elsewhere, creating a continuing spiral of change.
- Wicked problems are often not stable. Problem-solvers are forced to focus on a moving target.
- Wicked problems can have no single solution. Since there is no definitive stable problem or one right answer, there can be no definitive resolution.
- Wicked problems are socially complex. Their social complexity baffles many management approaches.
- Wicked problems rarely sit conveniently within any one person, discipline or organization, making it difficult to position responsibility.
- Wicked problems arise in a specific space and time, and so resolutions of one wicked problem cannot be transferred to another, although the learning can be shared.

When? Wicked problems can arise in any field of human endeavour. Any program of transformational change is, by definition, tackling a wicked problem since it involves social change. The collective learning spiral is specifically designed to address wicked problems.

See: Brown et al. (2010).

Xing the minefield

Why? Transformational change can seem a challenging and dangerous task to which different people adopt very different approaches. Different activities offer useful tools for helping people of different temperaments and ways of knowing to understand those differences and to appreciate the opportunities they present in working collectively to address a complex or wicked problem. Crossing a minefield is one such tool.

What? When a diverse group of people has come together to address a complex problem, it may be useful to ask each of them in turn to say how they would go about 'crossing a minefield'. This can be done either as a brainstorming session eliciting spontaneous responses, or by systematically working around the room, but in either case, each participant should be given the opportunity and encouraged to contribute.

Used in this context, the crossing is a fun and unthreatening way to help people realize that there are many solutions to challenging issues – and to become aware of how differently their colleagues may address complex problems when they arise.

Some of the responses to this exercise have been:

Very carefully
By hiring a helicopter
After mapping the area beforehand
Making a dash for it
Letting someone else go first

When? The exercise is useful at various stages throughout the collective learning cycle when people need to get to know each other better. The individual approaches above reveal strategies of caution, technical fix, selfcentredness, risk avoidance and risk-taking.

This is a particularly useful exercise to introduce at the beginning of Stage 3 in the cycle, when participants are seeking creative new ways of addressing the problem. It also allows the facilitator to understand the individual problem-solving styles.

See: Local Sustainability Project 1990–2010.

Yarning

Why? Sharing stories around a campfire or in another relaxed setting is a powerful way of building understanding between people from different cultures, generations or life experiences. As the Yarning Circle highlights in its program, 'The Yarn seeks to enhance our communication skills, leadership qualities and the strength of our relationships.'

What? Yarning is a way of communicating that comes from traditional Aboriginal practice. Sharing stories around the campfire with families and friends is central to the oral history of many indigenous cultures. It provides a place for voices old and young to be shared and heard.

Whether organized story-telling, or presented as a rambling discourse, yarning is the informal discourse through which people share many of their experiences. Like the yarn that is prepared for weaving or knitting, the spoken yarn contributed by each participant can be woven into a new story.

In collective learning groups where the members are already known to each other or where conflict based on past experiences is unlikely, informal yarning is potentially a very useful way for people from diverse backgrounds to get to know each other better and enhance understanding of each other's values and the goals that might bind them together.

When? Yarning can be used at various stages in a collective learning cycle. Once the initial setting of ideals is done, it can be useful in building shared understanding and mutual respect in a group whose members are not well known to each other. Yarning can be used in developing mutual trust by encouraging people to mix and yarn with someone they do not know, or someone from a different 'knowledge culture'. Equally, some yarning time may provide welcome relief and a more informal exchange of ideas in an intensely structured program seeking to address a particularly challenging problem.

See: Bennet-McLean (2000).

Zany ideas

Why? 'Blue sky' thinking and the imagining that pushes beyond (or floats above) traditional boundaries are often seen as contrary to sound objective science. But, at the same time, creative new ways of seeing complex problems are often necessary to go beyond where we are now and address those problems.

What? Brainstorming is perhaps the best known and most useful way of assisting a collective learning group to come up with creative new ideas.

A spark, such as a challenge, a new experience or a shock, can bring a spontaneous, rapid-fire series of ideas. The process proceeds without discussion or judgement. Contributors are encouraged to be creative in their thinking and do not have to justify their ideas. This is where trust and the use of dialogue are essential. All of the ideas are recorded as closely as possible to the original and are treated as of equal value. The brainstorm can be spoken, written, or pictured.

Only when the outpouring of ideas generated by the brainstorming process has been completed does the group pause for clarification of any ideas that are not well understood. Even then the zaniest ideas are not judged as good or bad.

- Most Nobel Prizes have been awarded to people whose ideas were at first dismissed.
- The first planes to go through the sound barrier crashed, until a test pilot thought of reversing the controls.
- An eminent scientist, when asked for a sophisticated technical test of life on Mars, replied, 'Just look for deliberate patterns.'
- A geographer who looked at the shape of the continents on the planet and said 'They are still separating from one land mass' was laughed at for decades, until the theory of continental drift was accepted as fact.

When? Brainstorming 'blue sky' ideas is useful whenever the group needs to move to fresh ground. Although it may need to be refined later in the collective learning cycle, no idea should be dismissed as 'zany' during the process. Any wild idea can lead to the greater creativity, insight, vision and originality that are needed to address a wicked problem.

Bibliography

Adam, B., Val Loon, J. and Beck, U. (2000) *The Risk Society and Beyond*, Cambridge: Cambridge University Press.

Adams, D. (2004) *Hitchhiker's Guide to the Galaxy*, London: Pan-Macmillan.

Adams, R. (1975) *Watership Down*, London: Macmillan.

Alexander, C. (2002) *The Nature of Order. Book one: The Phenomenon of Life*, Berkeley: The Centre for Environmental Structure.

Alliancing Association (n.d.) http://www.alliancingassociation.org/index [accessed 18.1.12].

Armitage, D., Berkes, F. and Doubleday, N. (eds) (2007) *Adaptive Co-Management: Collaboration, Learning and Multi-Level Governance*, Vancouver: UBC Press.

Arnstein, S.D. (1969), 'A ladder of citizen participation', *Journal of American Institute of Planners*, 35(4): 216–224.

Art of Hosting (n.d.) http://www.artofhosting.org/home/ [accessed 18.1.12].

Art of Moving (2006) Project hosted by Canberra Environment and Sustainability Centre, Australian Capital Territory, project leader John Reid, Canberra: Fenner School of Environment & Society, The Australian National University.

Bennet-McLean, D. (2000) *The Yarning Circle*, http://www.rilc.uts.edu.au [accessed 18.1.12].

Bohm, D. (1996) *On Dialogue*, London: Routledge.

Brian, D. (1995) *The Voice of Genius*, New York: Perseus.

Brown, L. (2005) *Vital Signs: The Trends That Are Shaping Our Future*, Washington D.C.: Worldwatch Institute.

Brown, V.A. (2005) 'Knowing: linking the knowledge cultures of sustainability and health'. In Brown, V.A., Grootjans, J., Ritchie, J., Townsend, M. and Verrinder, G. (eds) *Sustainability and Health: Supporting Global Ecological Integrity in Public Health*, London: Earthscan, Sydney: Allen and Unwin, pp. 131-163.

Brown, V.A. (2008) *Leonardo's Vision: A Guide to Collective Thinking and Action. Sustainability and Health: supporting global ecological integrity in public health*, Rotterdam: Sense.

Brown, V.A. (2011) 'Multiple knowledges, multiple languages: are the limits of my language the limits of my world?' *Knowledge Management for Development Journal*, 6(2): 120-131.

Brown, V., Grootjans, J., Ritchie, J., Townsend, M. and Verrinder, G. (eds) (2005) *Sustainability and Health: Supporting Global Ecological Integrity in Public Health*, Sydney: Allen and Unwin, London: Earthscan.

Brown, V.A., Harris, J. and Russell, J.Y. (eds) (2010) *Tackling Wicked Problems through the Transdisciplinary Imagination*, London: Earthscan.

Buber, M. and Smith, R.G. (1999) (1937) *I and Thou*, Edinburgh: T. and T. Clark.

Chapman, D.M. and Calhoun, J.G. (2006). 'Validation of learning style measures: Implications for medical education practice'. *Medical Education*, 40: 576–583.

Civic Engagement Booklet Series (n.d.)
http://asmallgroup.ning.com/ [accessed 22.1.12].

Coffee, J.C. (2009) *Gatekeepers: The Role of the Professions and Corporate Governance*, Clarendon Lectures in Management Studies, New York: Oxford University Press.

Dalkey, N.C. (1969) *The Delphi Method: An Experimental Study of Group Opinion*, prepared for United States Air Force Project, Santa Monica: Rand.

De Bono, E. (1999) *Six Thinking Hats*, New York: Back Bay Books.

Dewey, J. (1910) *How We Think*, Boston: Heath.

Donovan, M.S., Bransford, J.D. and Pellegrino, J.W. (eds) (1999) *How People Learn: Bridging Research and Practice*, Washington D.C.; National Academy Press.

Elix, J. and Lambert, J. (2007) 'Mapping the values of shorebird habitat in Tasmania: a tool for resolving land use conflict', *Conflict Resolution Quarterly*, 24: 469–484.

Fisher, R., Ury W. and Patton B. (2011) *Getting to Yes: Negotiating Agreement Without Giving In*. 3rd edn, London: Random House.

Freire, P. (1996) *Pedagogy of the Oppressed*, London: Penguin (orig. 1970).

Gardner, H. (1983) *Frames of Mind: The Theory of Multiple Intelligences*, New York: Basic Books.

Genuine Progress Indicators: Danaher, K. (2002) 'A better way to measure the economy', in Danaher,K. (ed.), *Corporations are Gonna Get Your Mama*, Monroe, Maine: Common Courage Press.

Hamilton, C. (1997) *Genuine Progress Indicator: A New Index of Changes in Wellbeing in Australia*, Canberra: The Australia Institute, Discussion Paper No. 14.

Holling, C.S. (1978) *Adaptive Environmental Assessment and Management*, Chichester: Wiley.

Huxley, A. (1932) *Brave New World*, London: Chatto and Windus.

Jung, C. (1964) *Man and His Symbols*, New York: Doubleday.

Keen, M., Brown, V.A. and Dyball, R. (eds) (2005) *Social Learning in Environmental Management*, London: Earthscan.

Keirsey, D. (1998) *Please Understand Me II: Temperament, Character, Intelligence*, Del Mar, California: Prometheus Nemesis.

Kelly, G.A. (1963) *Theory of Personality: The Psychology of Personal Constructs*, New York: W.W. Norton and Company.

Kilkenny, S. (2011) *The Complete Guide to Successful Event Planning* (with Companion CD-ROM), Ocala, Florida: Atlantic Publishing Company.

Knowles, M.S. (1980) *The Modern Practice of Adult Education: From Pedagogy to Andragogy*, Englewood Cliffs: Prentice-Hall/Cambridge.

Koestler, A. (1990) *The Act of Creation*, London: Penguin (Arkana).

Kolb, D.A. (1984) *Experiential Learning: Experience as the Source of Learning and Development*, Edgewood Cliffs: Prentice-Hall.

Kolb, D.A., Rubin, I.M. and McIntyre, J.M. (1974) *Organizational Psychology: An Experiential Approach*, California: Prentice-Hall.

Kolb, D.A., Lublin, S. and Spoth, J. (1986) 'Strategic management development: using experiential learning theory to assess and develop managerial competencies', *Journal of Management Development*, (5)3: 13–24.

Kuhn, T. (1970) *The Structure of Scientific Revolutions*, Chicago: University of Chicago Press.

Ledwith, M. and Springett, J. (2009) *Participatory Practice: Community-based Action for Transformative Change*, Bristol: Polity Press.

Le Borgne, E., Brown, V.A., and Hearn, S. (2011) 'Monitoring and evaluating development as a knowledge ecology: ideas for new collective practices', Amsterdam: IKM Working Paper No. 13.

Lewin K. (1943). 'Defining the "Field at a Given Time"', *Psychological Review*, 50: 292–310. Republished in *Resolving Social Conflicts & Field Theory in Social Science*, Washington, D.C.: American Psychological Association, 1997.

Local Sustainability Project (1990–2012) Canberra: Fenner School of Environment and Society, The Australian National University.

Lomas, J. (2007) 'The in-between world of knowledge brokering', *British Medical Journal (Clinical research ed.)*, 334 (7585): 129–132.

Luria, A.R. and Solotaroff, L. (1987) *The Man with a Shattered World: The History of a Brain Wound*, Boston: Harvard University Press.

Makridakis, S.G. (1997) *Forecasting: Methods and Applications*, New York: Wiley/Hamilton.

Mencken, H.L. (1921) *Prejudices: The Second Series*, London: Jonathan Cape.

Myer-Briggs Test (n.d.) http://www.personalitypathways.com/type_inventory.html [accessed 25.1.12].

Owen, H. (2008) *Open Space Technology: A User's Guide* 3rd edn, San Francisco: Berrett-Koehler.

Passmore, J. (1974) *Man's Responsibility for Nature*, New York: Scribner.

Peavey, F. (n.d.) *Strategic Questioning Manual*. http://www.thechangeagency.org/_dbase_upl//strat_questioning_man.pdf [accessed 18.1.12].

Piaget, J. (1951) *The Child's Conception of the World*, London: Routledge, Kegan Paul.

Popper, K. (1959) *The Logic of Scientific Discovery*, New York: Basic Books.

Rittel, H. and Webber, M. (1973) 'Dilemmas in a General Theory of Planning', *Policy Sciences*, 4: 155–169.

Robinson, L. (2002) Two decision tools for setting the appropriate level of public participation, http://www.media.socialchange.net.au/people/les/ [accessed 8.7.12].

Sandman, P.M. (1986) *Explaining Environmental Risk*, Washington, D.C.: U.S. Environmental Protection Agency, Office of Toxic Substances.

Sarkissian, W. (2010) *Creative Community Planning: Transformative Engagement Methods for Working at the Edge*, London: Routledge.

Schuler, D. (2001) 'Cultivating society's civic intelligence: patterns for a new "world brain"', *Information, Communication and Society*, 4(2) Summer.

Schuler, D. (2002) 'A pattern language for living communication: a global participatory project', In PDC '02 *Participatory Design Conference proceedings*, Palo Alto, California: Computer Professionals for Social Responsibility (CPSR).

Schuler, D. (2006) *Pattern Language for Civic Communication*, Boston: MIT Press.

Skolimowski, H. (1995) *The Participatory Mind*, London: Penguin (Arkana).

Smith, M.K. (2001). 'David A. Kolb on experiential learning', the encyclopedia of informal education. http://www.infed.org/b-explrn.htm [accessed 30.12.11].

Smithson, M. (1989) 'The changing nature of ignorance', and 'Managing in an age of ignorance', in Handmer, J. (ed.) *New Perspectives on Uncertainty and Risk*, Canberra: Centre for Resource and Environmental Studies, The Australian National University, pp 5-66.

Steinberg, D.M. (2004) *The Mutual-Aid Approach to Working with Groups: Helping People Help One Another*, London: Haworth Press.

Transition Towns Movement (n.d.) http://www.transitionnetwork.org/support/what-transition-initiative [accessed 25.1.12].

U.S. National Research Council (1999) *Our Common Journey: A Transition towards Sustainability*, Washington D.C.: National Academy Press.

Waldman, D. and Kuspit, K. (1993) *Studies in the Fine Arts, Criticism No. 28*, New York: DaCapo Press.

Walker, B.H. and Salt, D.A. (2006) *Resilience Thinking: Sustaining Ecosystems in a Changing World*, Washington D.C.: Island Press.

Wenger, E. (1998) *Communities of Practice: Learning, Meaning, and Identity*, Cambridge: Cambridge University Press.

Wondollek, J. and Yaffee, S. (2000) *Making Collaboration Work: Lessons from Innovation in Natural Resource Management*, Washington D.C.: Island Press.

World café (n.d.) http://www.theworldcafecommunity.org/ [accessed 18.1.12].

Yarning Circle (n.d.) http://www.theyarningcircle.com/ [accessed 25.1.12].

Index

Page numbers in italics = figure, bold = table,
'b' before the page number = information in a box

accommodation (learning style) 10, *10*, 51, 55

Action for Sustainability and Health 98–106, *99*, 105

action plans
  adaptive management *229*
  creating 63–4
  implementation 67–9
  negotiation 251–2
  as part of learning cycle 34

ActWise 102, 103

adaptive management 229–30, *229*

administrators 45

affinity circle *143*

Alexander, Christopher 253

alliancing 230–1

Arnstein's ladder *236*

Art of Moving
  design project 93, 102, 104
  project postcards *26, 59, 103, 104, 105, 115, 215*

assimilation (learning style) 9, *10*, 55

Australian National University, Local Sustainability Project 12–13, 17–19

balancing the players 231–2, *232*

Bali, Montessori school 193–204, b195–6, *199–204*

Beaconsville, Australia 107–14, b108

blue sky ideas 269

Bohm, David 240–1

Bono, Edward de 249

books, collective authorship
  connecting frameworks 175
  Social Learning and Environmental Management 185–92, *188*, b189, **191**
  Sustainability and Health 176–84, *176*, b183–4
  use of collective learning cycle 180–4

brainstorming 269

Brown, Emeritus Professor Valerie A. 24

Buber, Martin 51, 80

Canberra, Australia
    Action for Sustainability and Health 98–106, *99, 105*
    Energy Futures 116–19

case studies
    Action for Sustainability and Health 98–106, *99, 105*
    background to 89, 92–3
    changing problem communities 147–54
    collective thinking 155–64
    community change 97–114
    Community-wide Sustainability Services 107–14
    Elizabethville (polluted town) 148–51
    Energy Futures 116–19
    Ethiopia workshop 205–13, *206*
    Indigenous thinking (Australian) 161–3
    Integrated Knowledge Management 166–70, *168*, b169
    Integrated Sustainability Reporting 170–4, *172, 173, 174*
    Integrated Urban Alliance 138–46, *140, 143*
    introducing new ideas 115–24
    lateral and linear thinking 157–60
    long-term change 125–46
    monitoring and evaluation 165–74
    Montessori School, Bali 193–204, b195–6, *199–204*
    from other cultures 193–214
    Rivermouth (dying desert town) 152–4
    Social Learning and Environmental Management 185–92, *188*, b189, **191**
    strategic planning (case studies) 193–204, b195–6, 205–13
    Strategy for Healthy People on a Healthy Planet 120–4
    Sustainability - Our Future 126–37
    Sustainability and Health 176–84, *176*, b183–4
    Sustainable State-of Environment regional reporting 170–4, *172, 173, 174*
    teamwork 175–92
    types of change **92**

chair (person) 45, 50

change
    attitudes to 76–7, *76, 77*
    blocks to 21–2
    and conflict 4–5
    supported 36
    welcoming 5–6

change management practice 26

cities (case studies)
    capital 98–106, *99*
    regional 107–14

climate change, collaborative action 215–16

collaboration 107–14, 233–4

collective authorship 176–84, *176*, b183–4, 185–92, b189, **191**

collective decisions (knowledge cultures) 46

collective facts 53–5, 219–20

collective ideals 49–52, *49*, 218–19

collective learning
   and action *226*
   as a celebration 89
   collaboration 233–4
   conflict resolution 234–5, *235*
   consultation 236–8, *236*
   four stages of 14–15, b17, *22, 31*
   as a framework 32
   knowledge brokers 45, 50, 248
   mandala of *43*
   non-suitability of 47
   process of 16–19, b16
   rules for 26–9, 37
   Rules of Dialogue 240–1
   shifting between stages 52, 55, 60
   six steps of 22, *22*, 33–5, 211–12
   testimonials of b58, b133
   and transformational change 12–15
   and wicked problems 74
   see also learning styles

collective learning cycle
   applied to research 80, *81*
   collective authorship 176–84, *176*
   Ethiopia workshop *206*
   forecasting 243
   four stages of 14–15, b17, *22, 31*
   as monitoring tool 170–4, *172*
   multiple knowledges *22*
   open space rules 252–3
   six steps of 22, *22*, 33–7, 211–12
   synergy and synthesis 258–9
   transdiciplinarity 261–2
   see also learning cycles; learning styles

collective learning process
   implementation from the Guidebook (Bali) 193–204, b195–6, *199–204*
   implementation from the Guidebook (Ethiopia) 205–13, *206*

collective learning spiral 9, 31–2, *31, 35*, 175

collective learning teams 147

collective learning tools *see* tools

collective thinking 77–81, *81*, 155–60, b158, b160, 161–3

collective thinking (case studies) 155–64

collective transformation 90–1, *90*

community
   development 13
   educators 36
   reaction to transformation 64–5

community members 40–2, *40, 41, 42, 43*, 44–7, *46, 47*

community of practice 142–4, *143*

community sustainability goals *173, 174*

Index   277

community-based long-term change (case studies)
    Action for Sustainability and Health 98–106, *99, 104, 105*
    Community-wide Sustainability Services 107–14
    Energy Futures 116–19
    Integrated Urban Alliance 138–46, *140, 143*
    Sustainability - Our Future 126–37

Community-wide Sustainability Services 107–14

conflict resolution 234–5, *235*

consultation 236–8, *236*

convergence (learning style) 9, *10*, 55

conversation 238–9

core questions 33

Council organization (Beaconsville) 111

creative thinking 98–106, *99*, 102, *105*

critical loyalty 190

cross-cultural programs 166–7

crossing (xing) the minefield 268

culture 18

cultures, non-Western (case studies) 193–214

da Vinci, Leonardo 13

debriefing 52

decision-makers 39–40, *40*, 91
    see also knowledge cultures

dialogue 28, 240–1

direct experience 7–8

divergence (learning style) 9, *10*, 55

diversity 29

documents 50

doing, as part of learning cycle 8, 34

economic concerns 21

educational collective (case study), Integrated Urban Alliance 138–46, *140*, 143

Einstein, Albert 21, 77, b78

Elizabethville 148–51

Energy Futures 116–19

Ethiopia 205–13, *206*

evaluations
   holistic 144
   personal **145**

Even a small addition can make a difference *115*

event management 241–2

experiential learning 7–8, *7*, 11–12, 71

experts *see* specialist advisors

facilitators
   creativity of 57, 73
   role of 45, 50, 71–2, 251, 259, 263
   see also guides

facts
   conflict resolution 234–5, *235*
   as part of learning cycle 33–4
   workshop example *203*

feelings, as part of learning cycle 8, 33

Fisher, R. *et al* (2011) 251

following on 35, 36, 67, 68, 221

forecasting 243

futuring 243

Gardner, Howard 51, 80, 249

gatekeepers 244

Gestalt psychology 79, *79*

ground-truthing 19

guided imagery 246

guides
   creativity of 73
   identification of world views 74–5
   and learning styles of the group 75–6, *76*
   problem solving 72–3, *72*
   role of 71–2, 91
   see also facilitators

holists 40–2, *40, 41, 42, 43*, 44–7, *46, 47*

hosting 245

human condition, and transformational change b6

ideals
- for collective learning alliance *140*
- developing 49–51, *49*
- imagining 246
- risk and risk-taking 257–8
- visioning *59*, 265–6

ideas
- adaptive management 229–30, *229*
- blue sky ideas 269
- imaginative 220, 246
- implementing 60, 95
- as part of learning cycle 34
- risk and risk-taking 257–8
- zany ideas 269

imaginative thinking 57–9, 59, 246

individual learning 7–12, 155–60, b158, b160
- *see also* learning cycles

individuals 40–2, *40, 41, 42, 43*, 44–7, *46, 47*

Integrated Knowledge Management 166–70, *168*, b169

Integrated Urban Alliance 138–46, *140, 143*

intentional transformation 12

interest groups 39–40, *40, 41, 82*, 91
- *see also* knowledge cultures

invitations 44

joining in 247

Keirsey, David 232, *232*

knowledge and learning 19, 84, *84*

knowledge brokers 45, 50, 248

knowledge cultures
- creating collaboration between 44–7, 91
- importance of 221, *221*
- individual informed access to 82–4, *84*
- knowledge culture grid 141
- knowledge forms (Western) 40–2, *40, 41, 42, 46, 47*
- multiple knowledges 22, 250–1
- needs and resources cross-referenced **142**
- as nested system *43*
- as networked system *46*

Knowles, Malcolm 13

Kolb, David
- actionable knowledge 15, b16
- collective learning 16–19
- direct experience 7–8

experiential learning 7
experiential learning 7–8, 11–12
individual learning 7–12, 13
learning cycle 7–9, 31–2, *31*, 60
learning styles 8–11, *10*, 249–50

Kuhn, Thomas 75

Lambert, Dr. Judith A. 24

language 44, 237, 253–4, 262–3

lateral and linear thinking (case studies) 157–60

lead up (program) 36

leadership 91

learning cycles
adaptations to 18–19
collective 14–15, 22, *22, 31*, 32, 33–7, 91
collective learning 26–9, 47
individual 7–12, 155–60, b158, b160
monitoring and evaluation 167–70, *168*, b169
what can be? 15, *63*
what could be? 15, 57–60, *57*
what is? 15, 53–5, *53*
what should be? 15, 49–52, *49*
see also collective learning; collective learning cycle

learning environment 71

learning journals 134

learning spiral
embedded (case studies) 126–37, 138–46, *140, 143*
illustration of three turns of *35*
from learning cycle 8–9, 13, 64
rules of 26–9

learning stages, fragmentation of b16, 17

learning strategies 78–9

learning styles
Kolb's learning styles 8–11, *10*, 249–50
within the learning cycle 51, 55, 58–9
limitations of 11–12
within occupations 8, 76, *76*
strategies 78–9

Local Sustainability Project (Australian National University) 12–13, 17–19

Luria, Alexander 77

meeting environment 45, 49–50

modes of synthesizing evidence, 79 *79*

monitoring and evaluation
frameworks (case studies) 165
holistic 144
Integrated Knowledge Management 166–70, *168*
learning cycle *168*
personal **145**
Sustainability State-of-Environment regional reporting 170–4, *172, 173, 174*

multiple knowledges *22*, 250–1

Myers-Briggs Type Indicator 11, 80, 232

On the line 141

open space technology 27, 252–3

opposition to change, 64-5 64–5

order 18

organizations 40–2, *40, 41, 42, 43,* 44–7, *46, 47*

paradoxes 73, 74

participants
ideals 50–1
questionnaire for *82*
roles 82–3
see also knowledge cultures

Passmore, John 74

pattern languages 253–4

personal goals 144, *145*

poem b133

pollution (lead) 148

problems 72, *72*

problem-solving games 254–5

public health, and sustainability 177

questioning 256

questions, core 33

questions, project focus 44, 50, 51, 94, 251

redirected social learning 5–6

reflection 18

relationships, team
    balancing the players 231–2, *232*
    conversation 238–9
    gatekeepers 244
    hosting 245
    joining in 247
    problem-solving games 254–5
    questioning 256
    team building 56, 220, 259–61
    understanding 68–9, 262–3

reporting 52, 69

rewards 45

risk and risk-taking 257–8

Rivermouth 152–4

Sally and Richard (example guides)
    backgrounds 25
    collective learning trial 38
    developing team experience 94–5
    forming the action plan 65
    from ideals to ideas 61
    learning journals 134
    a new approach 30
    pilot project 38
    preparation for the big task 48
    shared journey 216–17

Sandman, P.M. 257

seven ages of man b4

shared learning, enhanced (case studies)
    background to 115–16
    Energy Futures 116–19
    Strategy for Healthy People on a Healthy Planet 120–4

Smithson, Michael 78

social concerns 21

social learning 3–6, 8, *188*

Social Learning and Environmental Management 185–92, *188*, b189, **191**

society, and wicked problems 73

solar power (Beaconsville) 111

specialist advisors 40–2, *40, 41, 42, 43*, 44–7, *46, 47*

speed dating 68–9, 144

stakeholders 42

strategic planning (case studies)
    Montessori School, Bali 193–204, b195–6, *199–204*
    Strategy for Healthy People on a Healthy Planet 120–4

Strategy for Healthy People on a Healthy Planet 120–4

succession policies 125

summarisation 52

supported change 36

sustainability
    Action for Sustainability and Health 98–106, 99, *105*
    community sustainability goals *173, 174*
    Community-wide Sustainability Services *104*, 107–14, b108
    Integrated Urban Alliance 138–46, *140, 143*
    projects 135–6
    public health 177
    Sustainability - Our Future 126–37
    Sustainability and Health 176–84, *176*, b183–4
    Sustainability State-of-Environment regional reporting 170–4, *172, 173, 174*

Sustainability - Our Future 126–37

Sustainability and Health 176–84, *176*, b183–4

Sustainability Summit (Beaconsville) 109–10

Sustainable State-of Environment regional reporting 170–4, *172, 173, 174*

synergy and synthesis 258–9

tasks, and collective learning 74

team building 56, 220, 259–61

teamwork (case studies) 175–92

technical solutions 177

Think journey Not destination 26

thinking Indigenous (Australian) 155, 161–3
    lateral 155, 157–8, b158
    linear 155, 159–60, b160
    as part of learning cycle 8, 34

timetables 36

timetables (change project) 36

tools
    adaptive management 229–30, *229*
    affinity circle *143*
    alliancing 230–1
    background to 44, 225–7
    balancing the players 231–2, *232*
    choosing 46, 225–7, *226*
    collaboration 233–4
    conflict resolution 234–5, *235*
    consultation 236–8, *236*
    conversation 238–9
    creative intervention 56
    dialogue 28, 240–1

284 Collective Learning for Transformational Change

event management 241–2
force field map 54, *54*
forecasting 243
gatekeepers 244
hosting 245
for ideals development 51
imagining 246
joining in 247
knowledge brokers 45, 50, 248
learning styles 249–50
On the line 141
multiple knowledges *22*, 250–1
negotiation 251–2
open space technology 27, 252–3
pattern languages 253–4
problem-solving games 254–5
questioning 256
risk and risk-taking 257–8
speed dating 68–9, 142, 144
synergy and synthesis 258–9
team building 56, 220, 259–61
transdiciplinarity 261–2
understanding 68–9, 262–3
value line 251
value lines b181
values mapping 263–4
visioning *59*, 265–6
wicked problems 21, 71–5, 250, 267
world cafe 69, 238
xing (crossing) the minefield 268
yarning 269
zany ideas 269

transdiciplinarity 261–2

transformational change blocks to 21–2; defined 3

Uluru, Australia *161*, *163*

understanding (through tools) 68–9, 262–3

users, potential 23

value lines b181

values mapping 263–4

visioning *59*, 265–6

walking school bus 112

watching, as part of learning cycle 8, 33–4

Wenger, Etienne 75

Westhaven Concerned Citizens 101, 103

Index **285**

what can be?
    adaptive management 229–30, *229*
    collective action 63–5
    Ethiopia workshop 211–12
    Montessori School, Bali *201, 204*
    negotiation 251–2
    stage 4 15, b17, 34, 63–4, *63*, 95

what could be?
    adaptive management 229–30, *229*
    Ethiopia workshop 210
    imagining 246
    Montessori School, Bali *199, 200, 202*
    risk and risk-taking 257–8
    stage 3 15, b17, 34, 57–60, 57, 95
    xing (crossing) the minefield 268

what is?
    conflict resolution 234–5, *235*
    Ethiopia workshop 209
    stage 2 15, b17, 33–4, 53–5, *53*, 95

what should be?
    Ethiopia workshop 207–8
    imagining 246
    Montessori School, Bali *199*
    risk and risk-taking 257–8
    stage 1 15, b17, 33, 49–51, *49*, 94
    values mapping 263–4

whole-of-community change (case studies)
    Elizabethville 148–51
    Rivermouth 152–4

wicked problems
    case studies (dying desert town) 152–4
    case studies (lead pollution) 148–51
    characteristics 72–3
    within communities 147–8
    defined 21, 267
    monitoring and evaluation of 165
    synergistic inquiry 258
    transformational change as 71–5
    types 250

workshops 36, 45, 51, 242

world cafe 69, 238

world views 74–5

xing (crossing) the minefield 268

yarning 269

zany ideas 269

An environmentally friendly book printed and bound in England by www.printondemand-worldwide.com

PEFC Certified
This product is from sustainably managed forests and controlled sources
PEFC
PEFC/16-33-415
www.pefc.org

FSC
www.fsc.org
MIX
Paper from responsible sources
FSC® C004959

This book is made entirely of sustainable materials; FSC paper for the cover and PEFC paper for the text pages.

#0384 - 160415 - C0 - 234/156/16 - PB